A review of the l
the practice
and the equi
by a 40-colony spa...
in diverse parts of England
during

Sixty Years
with
Bees

Donald Sims

FRICS NDB

About the Author

Born in 1914, Donald Sims has had bees of his own since the age of twelve, in 1926. Continuously since the mid 1950's he has had more than forty colonies, managed as a spare time activity of a busy man. He has kept bees in Kent and Sussex, in Derbyshire, in London suburbia, in Devon, in Northumberland, in Essex and Cambridgeshire, always all in out-apiaries. For more than thirty years he took twenty colonies or more to the heather moors each year. He has visited beekeepers both in Europe and in USA and western Canada. Throughout all these years he has selected queens to breed from and re-queened half his colonies each year. His colony management and queen selection and breeding practices largely derive from his sixty-year friendship with Brother Adam.

Listen patiently, and see whatever truth
other people's opinions may contain for you.
Think it possible that you may be mistaken.
Avoid hurtful and provocative language.
Do not allow the strength of your convictions
to betray you into making statements or allegations
that are unfair or untrue.

from Quaker Advices

For my children, Wendy, Nonie and Heather, who finally persuaded me to write this book, and for John and Phyl, and Colin, as promised.

Sims, Donald, Sixty Years with Bees

Published by Northern Bee Books, Scout Bottom Farm, Mytholmroyd, Hebden Bridge, West Yorkshire, England.

Line Drawings, Sketches and Photographs by the Author
Photograph of the Author by John Phipps

ISBN 0-907908-74-8 (PB)
ISBN 0-907908-79-9 (CB)

Printed in the UK by Watkiss Studios Limited, Biggleswade, Bedfordshire SG18 9ST

Sixty Years with Bees

This book is, in part, the story of one man's lifelong experience with bees; with forty colonies or more kept in diverse places in England; and of working alone and with another beekeeper or partner, moving colonies to orchards for pollination, to the heather, and to a variety of crops grown for seed.

It is also a book in which the author discusses bees and queens, hives and equipment, record keeping, out-apiaries, management practices and management systems, and writes clearly and thoroughly about swarm control and handling and marketing the crop. Helpful line drawings and diagrams by the author illustrate much of the text.

It is a book that the author himself describes as "a review of the lessons learned, the practices adopted, and the equipment used by a 40-colony spare time beekeeper in diverse parts of England during sixty years with bees."

Contents

List of Colour Plates

List of Line Drawings

List of Diagrams

Beekeeping is a science, a vocation, and a craft.
Some, having the vocation, attain a measure of proficiency
in the craft without designedly studying the science.
Others concentrate on the science, hoping thereby to attain to
craftsmanship,
but they seem to lack that touch of art or sympathy which clothes
the bare facts with life.

H.J. Wadey, "The Bee Craftsman".

Preface

Thirty-five years ago I was invited to write a book on bees and beekeeping, as a guide to those who aspired to run, as I then did, a large number of colonies as a single handed spare time operation. I was living in Devon, and had done so for some years, with forty colonies in two out-apiaries, and I sold my honey under my own label to a long established firm of bakers and confectioners in south London and my heather honey to Fortnum & Mason in Piccadilly. My spare time was limited by a full time and very demanding job, a large garden, and three children at an age when family outings to moor or sea were an important part of our life. I could not see how I could find time to write such a book, and I declined the offer. A few years later, when I was living and had my bees in Northumberland, I declined a second invitation for similar reasons.

Twenty years later I was urged by beekeeping friends to write such a book and to make time to do it. I have been fortunate to enjoy robust good health, without which my beekeeping would undoubtedly have been on a much reduced scale, but this book got started and was largely written while I was recovering from an illness, but for which I doubt if I should ever have found the time. I sent a nearly final draft to four of my beekeeping friends, and all four liked it a lot and were most helpful and encouraging. I got in touch with publishers. But then fate took a hand, and in spite of continued urging by my friends I have been unable to take it further until now. That has meant some revision and addition, to take account of the past ten years.

The background against which one writes today is very different from that of thirty-five years ago. At that time beekeeping was somewhat in the doldrums, with ageing beekeepers and few new recruits to the craft. Recent years have seen both a notable increase in the demand for honey and a re-awakening of interest in beekeeping. There is a far more open attitude to equipment and methods, and an evident keenness by those new to the craft to learn. Attend the BBKA Spring Convention at Stoneleigh and the situation is evident, and very encouraging, despite the threat of Varroa.

Varroa is fast becoming widespread. With the associated viruses of which we now learn, its effect has already been devastating. It poses a threat to beekeeping both as an absorbing and delightful hobby and to those who keep bees as a commercial activity. We already hear of winter losses such as to cause many beekeepers to give up beekeeping. We know that this happened elsewhere in Europe when Varroa arrived. But we also know that the number of beekeepers and the number of colonies have returned to former levels, or nearly so, in other European countries, and can surely expect that to happen in the UK too. We will have to learn to live with it, as others have done, and I don`t doubt that we will.

I took my first swarm in the summer of 1926, when I was nearly twelve years old, and I was given a Langstroth hive to put it in. Bees did not figure very largely in my life in those early years, although I had a colony or two from 1926 onwards and as many as eight in the early thirties. But I confess that bees have occupied a great deal of my time - of my spare time - for the past fifty years.

My other major interest outside my work is gardening, an interest shared with my wife, and the two interests have much in common. Gertrude Jekyll, writing of gardening, might have been writing of beekeeping when she said: "It is not easy at all. It has taken me half a lifetime merely to find out what is best worth doing, and a good slice of another half to puzzle out the ways of doing it."

My job post war required me to live in diverse places; in Kent; in Derbyshire; and in London suburbia; each for two or three years; in Devon for eight years; in Northumberland for nine years; and near Cambridge since 1971. For the whole of the time I've kept bees; for all but the first three years more than twenty colonies, and around forty colonies, all in out-apiaries, for all but a few of the following forty years. And that as a spare time activity in evenings and at week-ends, outside the requirements of a demanding full time job. Now, in retirement and with the help of a partner, he and I run sixty colonies..

Beekeeping is hard work, and is particularly so when working alone. But I have always thought it made sense to get the fullest return from one's effort, and to achieve job satisfaction in beekeeping as in other pursuits. My aim has been, and still is, not to keep the maximum number of colonies that I could manage in the time available, but to get the maximum honey crop for the amount of labour expended, and this is by no means the same thing. The time has always been severely limited by the requirements of other activities, to some extent flexible, but not expandable. The bees have necessarily been in out-apiaries.

One consequence has been a severely critical attitude to equipment and methods, and a periodical review of both with a view to improvement. I doubt if anything I have or anything I do is original, or even very unusual, but it is the result of trying and testing other beekeepers' ideas and suggestions. One inevitably learns a good deal from experience, but my principal debt is to other beekeepers, and particularly to those who have written for the benefit of others. I write now in an effort to repay that debt, and with the same intent.

There is at least one first class book available from libraries, if not at the booksellers, on every aspect of beekeeping, and I have appended a suggested reading list of such books so that my readers can read and learn from them as I have done. I shall write little about the anatomy and life cycle and instinctive behaviour of the honeybee, although a basic understanding of these is essential to success - indeed it is the foundation of good beekeeping practice. I write little about bee diseases, or the chemistry of honey. I do not attempt to summarise what is better studied in full. On specialist subjects it is best to read what specialists have to say.

I have been very reluctant to add to the number of books about bees and beekeeping and have twice declined invitations to do so. I doubt if I can add anything new, but I am finally persuaded that it would be useful and interesting to many beekeepers to attempt to distil from an unusually long and varied experience with bees the resulting practices and conclusions.

I write almost wholly from my own experience, and if my practices and views coincide with those of others, as they do, it is because we have each reached similar conclusions. I have bought, read, and have by me most of the books on bees and beekeeping published in English since 1940, and a number published much earlier. From some I have learned a great deal, and in writing now I have used some freely to check statements or to compare practices and views, but I have not consciously taken material from any except where I quote or acknowledge in the text.

The opinions I express are my own, and I try either to give my reasons for holding them or the background of experience and contact with others that has led me to hold them. My purpose in including autobiographical material has been both to enable the reader to read of the basis on which some of my practices have been founded and conclusions reached, and to permit me to pay tribute to those from whom I have learned, but I naturally hope that it will also be of some interest in reminding older beekeepers of times past and in giving newer beekeepers a little of their flavour.

In Gilbert White's words: "My remarks are the result of many years observation; and are, I trust, true in the whole, though I do not pretend to say that they are perfectly void of mistake, or that a more nice observer might not make many additions, since subjects of this kind are inexhaustible."

Much of what I have to say has been written up from lecture notes, and from letters, articles, and essays that I have written for publication in past years. I am grateful to the Editors of the bee journals for their permission to include some material that has appeared before. I must also thank those who so kindly read my drafts, and made so many suggestions. The book is the better for their attention. If I do not name them it is solely to avoid blame for the many shortcomings falling upon any but myself.

Debts accumulate as one writes a book. My principal debt is to my late wife, Joan, without whose interest, patience and forbearance, and occasional help, I should neither have kept bees all these years, nor been able to write this book. No beekeeper herself, for almost fifty years she was an unfailing source of wise counsel and cautious encouragement, and a responsive sounding board for ideas. Sadly, she died in the Spring of 1991. Despite urgent entreaties from those who read the earlier draft, and from other beekeeping friends, I have not felt able to take the book through to publication until now, and for that I apologise.

1   Retrospect

Looking back, I count it happy fortune
that cast a young man's lot among these scenes
and these people.

Q, "Memories & Opinions" (1944)

Pre-war

I have been interested in bees and beekeeping - some would say "hooked" - since I was about nine years old. I can't think why. My father never kept bees, and nor did any relations, as far as I know. But my father did play chess rather well, and played chess regularly with a beekeeper who lived near us in Kent, and it was through him, and through books that he lent me, that I first became "hooked". I recall cycling to his house to attend bee talks during the winter of 1925/26, when I was eleven years old, and going to the Saturday afternoon demonstrations in his garden the following summer.

At this time beekeepers were re-stocking with bees after the nearly complete extinction of bee colonies by the so-called Isle of Wight disease, and I particularly recall George Judge demonstrating at The Daisies and urging the adoption of modern methods and modern hives.[1] George Judge was a great man in Kent beekeeping at that time, and an ardent advocate of the Langstroth hive. Arthur Dines, a friend of long standing, knew George Judge well, and had his bees in Langstroth hives, and initially so did I.[2]

I spent much of my boyhood in Romney Marsh and got to know a beekeeper in Lydd named Barnes, whose sideboard displayed the cups and bowls that he had won with his honey as his bee shed did with his prize cards.[3] He used to take me with him as his "smoker boy" (but mainly to fetch and carry, and for company) and used to talk through all his bee operations much as the advanced driver is required to do. What a way for a youngster to learn!

Barnes' bees were in the midst of a vast area of shingle - the Dungeness beach - with little or no soil and not much vegetation. A most unlikely place for honey production. But nectar poured in, from birdsfoot trefoil and the like in early summer, and later from vipers bugloss (Echium vulgare) which hummed with bees in its marvellous blue haze from June to September.

Comb honey was Barnes' speciality, and years later, when I read Carl Killion's outstanding book "Honey in the Comb" it immediately struck me how similar Barnes' methods had been. Barnes later gave me a Woodman Bee Smoker that I still use today and would be lost without - but of course most of the parts have been renewed from time to time.

Through chess my father also knew Leonard Illingworth, who lived in the village of Foxton, near Cambridge, where I now live myself, and I met him once or twice when my father called on him, and in later years.[4] Both Illingworth and my father were regular participants at the Hastings Chess Congress. Illingworth was very actively involved with "Bee World" throughout its early years. He was a prolific writer on bees, with international contacts, and in considerable demand as a lecturer. He had 60 colonies of bees in his one apiary. Old residents told me that they did well; and he was not the only beekeeper in the village. Since 1971 I have lived within a mile of Illingworth's house at Foxton and can say that it is today some of the poorest bee country in England. There are four colonies of bees in the parish (mine) and they barely survive. Then there were pastures and white clover; now there is nothing but tillage, growing wheat, barley, sugar beet and potatoes, and sometimes beans. There is oil seed rape within a mile or two but none in Foxton parish.

Through Illingworth, I met Gauntlet Thomas,[5] who kept bees near Newmarket in Simmins' Commercial (16x10) hives[6] and seemed to produce prodigious quantities of honey, as I recall, much of it from sainfoin. There is no sainfoin today, or none that I know of. I took eight colonies to sainfoin at Barton, near Cambridge, in 1974, to what I believe was the last field of sainfoin grown. The bright yellow wax and the delectable honey is quite unforgettable - but alas no more to be had.

I called on Gauntlet Thomas once or twice in the 30's when I was living in Suffolk. He wrote to me a year or two after the war, when he was giving up beekeeping, and I bought a few colonies from him. His bees were very good, but I did not like his hive, a distinctly heavy thickly painted affair with bottom bee space and rather flimsy frames for their size. I should add that I think well of the 16x10 frame and Commercial hive providing it has top bee space. These didn't.

I have had bees of my own since 1926. In the 1930's I was learning to farm, first in Kent, and later in Suffolk. For three of these years (1933,34,35) in Suffolk, I had eight colonies of bees in Langstroth hives. I met Brother Adam for the first time at a bee meeting at Rothamsted in 1935, at which he was one of the speakers. He was then on the right side of forty, and I was nearly twenty-one. It was there that I first heard him refer to bees that were "reluctant to swarm". Suffolk was very good bee country in the 30's. A local beekeeper (Jim Fisk) and I worked our bees together. I have a photograph showing our hives in 1934 higher than we could reach and topped with makeshift supers (without frames, as I recall, because we had no more).[7]

The bottom had gone out of farming and bees thrived on the neglect. Thistles and charlock flourished in arable fields with brambles in uncut hawthorn hedges yards wide.[8] I recall also that there was no sale for the honey which we got. By the end of the 1934 season we had more than a ton of honey in ten gallon milk churns, and the best offer we had for it was seven pence a pound. I think it was fully three years later that the last of it was sold, (for a shilling a pound) by which time I was living in Herefordshire, on the Welsh border, and had sold my bees.

Post-war - to 1955

In the summer of 1944 I was stationed in a large requisitioned house with an orchard and grounds at Pinner, in Middlesex, I well remember being informed of the discovery of a swarm of bees in the orchard. "Like a great footba' it is, sorr...your instructions, sorr?" I asked for the police to be informed. The police advised my office to get in touch with Mr. Carey, the County Bee Adviser. He called the next day to discover the swarm comfortably housed in a grocers box or whatever on the grass beneath the branch on which it had clustered, and when he found that it was I who had done this thing I could not persuade him to remove it. In no time at all a very good WBC hive stood in the orchard and housed the swarm. Protestations were useless, no heed paid to the uncertainty of one's domicile or future, so I bought it (for temporary ownership, I thought) and have not been without bees since.

The following year I was back in my native county of Kent and by 1947 I had sixteen colonies in the orchard country I knew so well. But do remember that I was not a beginner with bees. I had bought National and the then new Modified National hives, but I had already decided to adopt the Smith hive so that increase was made into Smiths. Since 1949 I have used Smith hives exclusively.

In the years just after the war I visited a great many people in the rural areas of southern England in the course of my work. Many of them were farmers, and some of them kept bees. Two of these used the equipment and methods of an earlier generation. In Blean Woods, north of Canterbury, an old woodman had fourteen colonies in straw skeps, with hats, on stands. His methods were those of the traditional skeppist. One day I watched him drive bees (from one skep to another, for sale). I found two Simmins' Conqueror hives in use at Hartfield, in Sussex, and was shewn how this remarkable piece of equipment was designed to work; with deep and shallow boxes like drawers in a chest, to be slid out and in on runners. That was not how the beekeeper used them; they were stuck fast with propolis. I thought that inevitable, and he couldn't remember them being otherwise. Looking back, I wish now that I had bought them, as I could have done, and given them to a museum for posterity to see. They were unique - and quite impracticable.

The years 1945 to 1955 were exciting bee years for me. In 1946 I had met John Hunt, at, of all places, a Saturday afternoon demonstration at The Daisies, which was still the venue for the local Bee Association meetings, almost twenty years after I had first attended bee meetings there. John and I found an immediate rapport (which we still have) and before long were working together with our bees. John was a dessert apple grower (one of the best) and my sixteen hives were in an orchard about a mile from his house.[9] A mutual friend, J.L. Leibenrood, who had a similar number of colonies in National hives, soon joined with us in what became a joint enterprise which we ran together for eight years. Between us we ran fifty to sixty colonies.

Through J.L.'s American connections we were able to import a number of American Italian queens from several of the breeders in the southern States. I recall that we had them from Daniels, from Weaver, and from Calvert, among others. They all had an infinite capacity for breeding the most beautiful and most docile bees you ever saw. The Daniels bees were a pale, pale gold, and almost transparent. These bees were busy and industrious, but they turned all their energies to producing bees, and it was soon evident that they were useless as honey producers in our climate. Their one really valuable characteristic was an astonishing willingness and ability to draw the most perfect combs from foundation, and this I soon set the colonies to do. Twenty years later I still had a few combs in use that I could identify as having been drawn by those bees - and my standard for culling combs is pretty high.

Colonies headed by daughters of these queens, mated locally, perhaps to our Buckfast drones, were a very different cup of tea. Outstanding as honey producers (our first ever 150 lbs. surplus from one of them), a bit inclined to swarming, with a quite nasty temper at times and at best unreliable. Quite unsuitable to breed from we thought - and we didn't.

About this time we met my namesake, but no relation of mine, R.P.Sims, who had moved his honey production and queen breeding activities into the Canterbury area.[10] In those days we could get good English reared and mated queens from a number of breeders, Claridge in Essex, Roberts in the Isle of Wight, R.P. Sims in Kent, and others. Those that we had from Claridge and Sims reminded me of the very first queen I bought, a Fairways Honey Queen, for 7/6d, from Capt. Turner at Ightham, in Kent, in 1927. I bought a Langstroth hive from him in the same year. New, in the flat, complete with Hoffman frames, floor, cover board and roof, it cost twenty-five shillings. It would cost ninety pounds today.

R.P. Sims mated his queens from mininucs - tiny boxes using one or two sections with starters as breeding frames and stocked with a pint or so of bees. John and I made some identical boxes and put them to use for our own queen rearing purposes, and although we had some success we had too many failures. The bees just wouldn't stay in the boxes for us, and we would find tiny swarms hanging around the place, so we abandoned mininucs. No doubt the fault lay with us. Mininucs are used much more successfully today.

In 1949 and again in 1950 and 1951 Brother Adam sent me a Buckfast breeder queen, and after testing the daughter queens we thereafter stuck as close to the Buckfast strain as we could.

As an apple grower, John had transport suitable for moving a load of bees, and for the seven or eight years we worked together we would put a truck load of about 20 colonies into a variety of seed crops in Romney Marsh and elsewhere in Kent and Sussex.[11] One of the first of these moves (to an 80 acre field of turnips grown for seed - very much like oil seed rape) taught us a never to be forgotten lesson, namely to see and know where the bees are to be put before you take them there.

The seed grower couldn't have been more helpful, and had told us that he had cleared a patch within the crop where the bees could get shelter right opposite a bridge across the surrounding dyke. What he did not say, or realise what difficulty it would create, was that the bridge was two planks wide, with no rail or guard, just above water level, and well down the grass banks of the dyke.[12] It was bad enough getting the colonies in, but just imagine the difficulty of getting really heavy colonies, with an average of 80 lbs. of turnip honey in the supers, out again.

That turnip honey was never beaten at any show to which it went - and I sent it to at least five of the major shows later that year - both in the light honey and in the granulated honey classes.

We took colonies to the New Forest for heather honey, with tolerable success. But it was not until later, when I lived for two years in Derbyshire, that I got to be "a heather man". Twice we got a useful crop of honey from red clover being grown for seed on the lower slopes of the South Downs, as my colonies near Lewes had done earlier.[13]

The orchard country of the Weald of Kent is, or was, very difficult beekeeping terrain. The main, and often the only, honey flow is from the orchards, cherries then apples, starting before the end of April and over before the end of May. Thereafter only minor sources of nectar, from holly, sweet chestnut and bramble in the woods. Colonies of bees hired by the fruit growers for pollination were brought into the orchards in large numbers and later taken home again. Beekeepers who lived in the area and did not move colonies out could expect little but pickings from hedgerows and woods, bramble towards the end of the summer, and on really warm days a minor flow from sweet chestnut, which produces a honey that looks just like engine oil and has a very strong flavour, which fortunately improves with keeping.

To take maximum advantage from the fruit blossom really powerful and well found colonies are required by the end of April. The only way this can be done, as far as I know, is to winter really powerful lots with good queens in large hives with bounteous stores. Such colonies, very different from the relatively small lots brought in for pollination, will commonly get one or two supers of apple blossom honey, and it makes obvious good sense to take such colonies to later sources of nectar.

My job took me to Nottingham for three years in 1950/52. I took four colonies of bees with me, and they found the pattern of the seasons there very much to their liking. My small out-apiary was precisely where the Trowell Service Area on the M1 motorway now is, and I think of it when I pass. I met a number of good beekeepers and interesting characters in the Derbyshire BKA, and here I first met Beowulf Cooper, the energetic and

dedicated founder of the BIBBA.[14] His death is a sad loss to
British beekeeping.

Stationed at the London HQ office from 1952 to 1955, I
was able for another three years to continue with the peripatetic
beekeeping that John Hunt, J.L. and I had developed. For eight
consecutive years we ran fifty to seventy colonies on a common
basis; queen rearing on behalf of the three of us so that we soon
had the same strain of bee; colonies belonging to each of us in
similar proportions in each location; regular inspections at
weekly or nine day intervals by any two of us (and sometimes by
all three); and inspection and operation without regard to
ownership. J.L. had 16 colonies in Modified National hives,
some with Hoffman frames; I had 20 increasing to 30 or so all in
Smith hives; John had half his colonies in Smith hives and half
in Modified Dadants and a similar total number to me. The
same bee, the same management, the same out-apiary sites and
wintering sites, with some colonies belonging to each of us at
every one, for seven full consecutive seasons and into the eighth.
Only the type of hive differed.

Our conclusions and results are therefore of some
interest. They were, first, that the advantages of truly single
walled hives, with top bee space and Hoffman frames, were soon
so evident that there could be no question about it. It always
took significantly longer to inspect colonies in National hives
than in Dadants or Smiths, and the bees were nearly always
less easy to handle. The need to release frames that "lift" with
the upper box and to free metal ends from the box and from each
other is both time consuming and disturbing to the bees.
Secondly, that the excellent handhold that the Modified
National hive provides has to be set against the added weight
and cost and significantly greater floor space in the truck when
compared to a Smith, even if the National had top bee space and
Hoffman frames. Third, that none of us would buy or use a
National hive by choice, but as between Smith and Dadant the
balance was not decisive. The Dadants consistently gave a little
more honey (up to 10% more) than a double Smith, but had no
advantage over a season in inspection time, had less flexibility,
and involved some really heavy lifting. Dadants had the
advantage of using less frames and combs and thus of less time
when making them up and fitting foundation and at extracting

time. Had we been certain that two man operation would always be available to us we would have given the vote (narrowly) to the Dadants. Actually we gave it to the Smiths. My vote goes the same way today.

Devon - 1955/62

In the early part of 1955 I moved to Exeter, and my wife and family joined me in August. During that summer I moved twenty-four colonies of bees from the orchard wintering site in Kent and established two out-apiaries within a few miles of Exeter. John and I made the journey by truck. By that time our bees were very close to the Buckfast strain. Ten of the twenty-four colonies were headed by daughters of one Buckfast breeder queen, and ten by daughters of another. They were equally split between the two apiaries.

1955 was a season to remember. We had taken off rather better than a super apiece of apple blossom honey before moving the bees to Devon. By the end of the summer I had taken an average of more than 200 lbs. per colony from the twenty-four colonies, and had a further 210 lbs. of heather honey from the six colonies taken to Dartmoor.[15] There was also an increase (by two) in the number of colonies.

I must add that the next two or three years were far from good, and I was glad to have 1955 honey in store to sell. Five sister queens topped the 1955 averages, and I bred from two of these queens in later years.

Fred Richards, who had been CBI in Devon before moving to Norfolk, visited Brother Adam with me, and gave me valuable advice on where bees might do well.[16] The best areas for bees are not to be found close to Exeter, and I moved the bees from one of my two out-apiaries further afield to Bow, beyond Crediton. It was twenty miles from home but proved to be a really good location and my bees did well. We got hawthorn honey there in several years.

George Jenner, who had succeeded Fred Richards as CBI for Devon, was also very helpful. He had adopted the 16x10 frame, and methods closely following Brother Adam's, mating queens and overwintering nuclei on Dartmoor and re-queening as a first spring operation. He used sometimes to give me a hand in my Bow apiary. He told me years later that his 3-comb nucs had come through the awful 1962/3 winter on Dartmoor quite unharmed, although they had been under deep drifts of snow for long periods.

Much of Devon is good bee country. There are a great many beekeepers, and a very active County Bee Association and branches. I found it a stimulating county to live and keep bees in, and made some good friends. My bees remained in two out-apiaries, one about 10 miles and the other about 20 miles from home, with a home queen rearing base and an annual migration of most of the colonies to the heather moors. There is surprisingly little heather on Dartmoor, and that only at the highest elevations and all too often sitting in a cloud. I had taken six colonies to the heather on Dartmoor in 1955 and I took some there each year until 1962, but I found Exmoor more rewarding and took rather more colonies there after 1956.[17] I had established a regular outlet for honey for sale before moving to Devon, and was fortunately able to expand the outlet through the same firm. I found a new outlet for heather honey, much of it in the comb.

These were years of a single handed operation of a thirty to forty colony outfit. My wife, and sometimes the children, would always give a hand with the move to the moors (although it meant a 4 a.m. start). A moorland breakfast after the colonies have been set down and released is a memorable occasion. I had a trailer made in 1956 to take ten hives (or twenty in a double deck). It has proved invaluable, and is still in regular use.

Northumberland 1962/71

Early in 1962 I was posted to Newcastle-upon-Tyne, and during the course of 1962 I moved the 40 colonies, supers, etc., from Devon to Northumberland, some in the late spring, and some straight to the moor in August. With help from Colin Weightman, whom I'd known for some years, I found apiary sites in the Tyne valley, near Corbridge, and I had bees there, or nearby at Hexham, until 1984.

The awful winter of 1962/63 was my first in Northumberland. We had heavy falls of snow in November, which had not fully gone by Christmas, when it started to snow again. We didn't see the earth beneath until well into April. Two cars and a motorbike were abandoned and remained buried in a snowdrift for six weeks within 200 yards of my home. The bees appeared to come through the winter well, nevertheless, and I lost only one colony at home and one in the Tyne valley, both queenless. But about a third of those that appeared to come through the winter well dwindled rather than built up and no more than a third made satisfactory progress and growth. It was a typical local experience, and we never did discover the cause or causes. Doubtless confinement without opportunity for flight for at least fifteen weeks had something to do with it.[18]

In much of Northumberland beekeeping would be a most unrewarding activity if it were not for the heather. The heather moors are vast, and within twenty miles of every beekeeper's doorstep, at a relatively low elevation of 700 to 1000 feet, and carrying almost nothing but ling heather. Small wonder that most Northumberland beekeepers regard the heather as the main crop and manage colonies accordingly. So therefore did I. The Tyne valley bees often got a useful flower honey crop too, but one didn't count on it and accepted it as a welcome bonus.[19] I found that Smith hives were very nearly the local standard equipment; certainly more numerous that Nationals or other types. Also that a short lugged 14 x 8½ inch frame had been in use in heather going hives for fifty years or more, and there were still some of these old hives about.

The last of the peripatetic annual agricultural shows of the RASE was held on Town Moor, Newcastle, in 1962,[20] since when all the RASE shows have been held at the permanent showground at Stoneleigh, where the BBKA Spring Conference is also held. I met Ernie Pope during the preparations for the 1962 Show, and we gradually came to work more and more closely together with our bees.

We helped each other, and made a joint move to and from the heather, for eighteen years. He usually accompanied me on my visits to my Tyne valley bees. We worked together in a remarkable harmony and unspoken understanding. His swift diagnosis "this one's allright", or the opposite, was rarely off beam. We learned a great deal from each other over the years, and enormously enjoyed working together. Almost as a bonus, we achieved a lot more together than either of us could have done alone.[21]

Quite unexpectedly, in the latter part of 1970 I was offered a job in Cambridge that I could not refuse, and took up the new job early in 1971. It meant even less time for beekeeping, and there was no choice but to reduce the number of colonies and to leave those on one out-apiary site in the Tyne valley for operation on a minimum visit basis. I have something to say, later, about operating twenty colonies some 250 miles from home, as I did for fourteen years.

We missed seeing bees at work in the garden, and in 1973, when work pressure had somewhat eased, I set up two colonies at Foxton, and later put some more on a farm in Essex, fifty miles from home, in a good area for bees, as I was told, and as it proved to be. The twenty colonies in Northumberland were brought down to join them in 1984, which made forty-four colonies there. A young man who came to help me in 1986 knowing nothing about bees is now my partner, and we run sixty colonies. Without his help no one can doubt that I would have greatly reduced the number by now had I been working alone.

Easier beekeeping

Like most beekeepers, I greatly enjoy working with bees. The necessary concentration on the job in hand when doing so excludes all thoughts of anything else whatever. I have no doubt of its therapeutic value. My wife used to say it kept me sane. But beekeeping is hard work, particularly so if working alone, and with the number of colonies I have had, and my aim has always been to make the work easier and to get a maximum return for my effort.

Mainly to increase my knowledge of bees, I studied and sat for and passed the really rather severe examinations for the National Diploma in Beekeeping, thirty years ago. One result was an increase in demand for my services as a lecturer, which I much regret that I then had often to refuse, as I was so fully occupied with my job.

Week-end schools at Askham Bryan, Bishop Burton, Newton Rigg and Seale Hayne, when I was able to accept, are memorable occasions. One meets many old friends, and makes new ones. Happily, I was also able to accept the invitation by the Scottish Beekeepers Association to be their visiting lecturer in 1977, and I gave a series of lectures in Kelso, Edinburgh, Dundee, Elgin, Thurso, Glasgow and Dumfries. Ted Hooper, and others, have been similarly honoured. It was a great experience. My lecture notes are a significant ingredient of this book, as are also my hive record books.

In a lecture that I gave some years ago, which I called "Easier Beekeeping", I listed seven points, as follows:

Get a partner - beekeeping is a two man job.
Bees - keep only the best.
Equipment - adopt and use only the most efficient, and standardise.
Apiaries - find good sites, and adopt an efficient arrangement of colonies.
Methods - simplify, clarify objectives, and learn to handle colonies speedily.
Nectar - keep bees where there's nectar to be had, moving if necessary.
Records - keep and study colony records.

The first point caused amusement, but it was seriously made and should never be overlooked. Operating with another beekeeper with whom you have or can develop an affinity in working not only adds greatly to the enjoyment and progress in learning (by both) but also more than halves the effort of each. There are many operations that are easy and quick when there are two people to carry them out, but are quite difficult and slow if one is on one's own.

The other six points are the subject matter of this book.

2   Bees and Queens

There is no gainsaying that the strain of bees
is the first and most important factor
whether bees are kept for pleasure or profit.

Brother Adam, "Beekeeping at Buckfast Abbey"

Bees

When I moved to Northumberland in 1962 Colin
Weightman had about 120 colonies in four or five groups around
the farm. They were wintered and run on two National size
brood boxes, and the power and strength of these colonies was
impressive. Colin is a very skilful breeder of bees (and of cattle
too, I might add) and his strain was derived from a combination
of the Buckfast bee with the local dark bee used by Robson and
Cessford in earlier years. It required both careful selection and
the maintenance of the original breeding stock in its pure state,
and he had an isolated queen mating apiary on moorland ten or
twelve miles away.[1]

Colin's remarkably powerful colonies were not difficult to
handle and not given overmuch to swarming, and really did get
honey, both in the valley and on the moor. With his
encouragement and help I set about changing my own strain,
then very close to Buckfast, to something close to his. I had
colonies in his apiaries and elsewhere on his farm, and worked
with Colin and he with me, which greatly helped.

My bees in Ponteland, 15 miles north, and in a later,
more exposed, area, took a rather different course. After the
dreadful 1962/63 winter I bought some package bees, among
them Caucasians of the Hasting strain, and these interested me
very much.[2] I got a breeder queen from J.E.Hasting, in northern

Saskatchewan, and persuaded Ernie Pope (who had local black bees of good character and performance) to try some of the daughters. It soon became apparent that these bees showed great promise. Ten years later, both Ernie Pope's bees, and my own, were a pretty uniform lot, developed from a combination of this strain of Caucasian bee with the local dark bee, and as good for these northern conditions as any we knew.

We were fortunate in being able to find, deep in the Border Forest, a mating station seven miles from any known bees, and there set up in a forest ride drone colonies selected from our best, to which we took Caucasian virgins in nucs, and the following year took Caucasian drone colonies there for further matings. We found to our surprise that the little three or four comb nucs not only maintained themselves, deep in the forest, but actually added to their stores, presumably from spruce honeydew, and it was very pleasing if not surprising to find the queens mating quickly and without loss. But oh, how the flies bit! Tiny flies that got through a bee veil. Flies which also plagued the bees, we thought. Years later, when I went to see Hasting and other beekeepers in Alberta and Saskatchewan, I found similar tiny flies as bad or worse in the Canadian forests.

Ernie and I were able to assess and choose from about 60 colonies each year, and we had a few clear guide lines, namely to breed queens deliberately and only from our chosen breeders, and preferably from more than one each year; to select drone breeding colonies; to eliminate queens from colonies that made swarm preparations in the year the queen was born or in the following year; and regular requeening after two full seasons, except for breeder queens.

One needs to have a clear aim and to keep it clearly in mind. You won't achieve it, but you will be aware of the area and extent of the shortcomings. The desirable characteristics that I seek are those which Brother Adam sets out in his "Beekeeping at Buckfast Abbey". Do read what he has to say. I have always tried, as he does, to test sister queens of the same production batch in different apiaries. Superiority in all apiaries is so much more conclusive than superiority in one. To minimise drifting I adopt the same arrangement of colonies in my apiaries as Brother Adam does at Buckfast. So does Colin Weightman. Colony records need to be as reliable as possible.

I knew Beowulf Cooper before BIBBA was formed, and I have a great admiration for the strides that BIBBA has made both in achieving combinations of like minded beekeepers and in exploring, testing and publicising queen rearing methods and other aids to better beekeeping.[3] There are undoubtedly some very good strains of local bee, well suited to the district, which it is good sense to preserve and develop. But the emphasis on "saving and popularising the best of the less prolific breeds closest to the native bees of the last century" is not one that appeals to me. For many years prior to 1870, when none but native British bees were to be had, Pettigrew, and his father before him. were getting bigger crops of honey than their contemporaries from colonies in straw skeps with a capacity similar to a Langstroth Jumbo or British Deep brood box.

My aim for forty years has been maximum production per colony with minimum labour, and my bees have consistently indicated that they need a brood chamber larger than a single National or Smith box, of the order of that plus a shallow, and happily accept and do very well in a still larger brood chamber (in my case two Smith boxes) which is never short of stores. Only with a brood chamber of that size have my queens been able to develop their full potential and their colonies to give maximum yields. I remain quite unconvinced that I, and my colonies, would have done as well had they been less prolific.

Because such a high proportion of beekeepers in the UK use the 14 x 8½ inch frame there is, I fear, a continuing tendency to find a bee that will be content with one box of such frames as a brood chamber, instead of unhesitatingly using a brood chamber adequate to really productive bees. It seems to be a wish to have bees to fit the box instead of a box to fit the bees.[4]

One should have a clear ideal in mind, and work towards it. I require bees that are easy to handle, that don't require much smoke and don't need constant smoking, are good tempered and don't pester or follow; bees that stay quietly on the combs and don't run or bunch; and a queen that carries on with her work and doesn't run or hide.

It is always helpful if the queen is seen. Whether she is seen or not depends far more on her behaviour than on any marked contrast in colour or colour marking. I don't colour mark my queens, and never have. In my record book the first note is always whether or not I saw the queen, and it is quite remarkable how some queens will be seen nearly every time and others almost never - these are the ones you have sometimes to hunt for with consequent waste of time - and I find this aspect of queen behaviour strongly hereditary.

Of course, I also aim to have bees that are not given to overmuch swarming, and I particularly dislike strains of bees that follow a swarm with numerous casts. I seek bees that winter economically, cleanly, and well. Most of mine do. Breeding only from the best, and never from those with obvious faults, culling those that disappoint, leads to uniformity at an acceptable level of performance and makes apiary management enormously easier.[5]

Brace comb (between boxes and between frames) is a nuisance and a handicap to ease of operation. Poorly designed frames and failure to maintain bee spaces is a principal cause, but so too is a strain of bee with a tendency to build it. Some strains of bee propolise everything, some use very little. Our Caucasian derived bees built barriers of propolis at the entrance, which sometimes stuck down the front corners of the frames, but used very little elsewhere. Bees that use propolis immoderately on and around the frames, particularly around the frame lugs, make operation unnecessarily difficult and unpleasant.[6]

Naturally, one breeds from queens heading colonies giving consistently better yields than others under similar conditions, and that can't be wrong, provided the selection also has full regard to the other desirable characteristics and to serious faults. I have always tried to keep the best bees I know, and that has meant buying queens from time to time to improve the strain, introducing bought queens and home-bred queens, and rearing queens.

Queen rearing

I suppose almost every beekeeper seeks to keep bees of a thoroughly desirable strain, well suited to his locality and circumstances, and exercises some degree of selection towards that end. It makes good sense, to choose which queen to breed queens from, and to rear these queens deliberately, and well. Good queens can be got by quite simple methods.

An abundance of pollen and at least a modest daily intake of nectar is necessary to the production of good queens, and queen rearing should be undertaken only when both can be expected to be available. Oil seed rape is a prolific source of pollen, and often of nectar also, and colonies at the rape commonly swarm, so I rear a first batch of queens at that time.

The simplest possible method - and an excellent one if only a few queens are needed - is to stimulate a selected colony wintered on a single box to build up rapidly in spring by feeding frequently and regularly, and to induce preparations for swarming by failing to provide room for colony development. Proceed as follows: In August or September select a colony with a good queen at the end of her second year from which you would be happy to breed daughter queens, and winter this colony in a single box in the home apiary. From early spring onwards encourage the development of this colony by feeding, if necessary, both pollen and syrup. Add no further boxes. Inspect for queen cells from time to time.

When advanced queen cells are found, and just before the colony would otherwise swarm, make an artificial swarm with the queen by moving the colony away and providing a new box with foundation and one empty comb on the old site, into which the queen and the bees from two or three combs are brushed or shaken, and joined by the flying bees. It will pay to feed this artificial swarm.

Three or four days later the parent colony can be split into nuclei, each with a queen cell, and some three weeks later each should have a young laying queen.

Without drawing upon other colonies for brood and stores, however, this method will probably not provide more than four young queens, as more nuclei than this cannot satisfactorily be made.

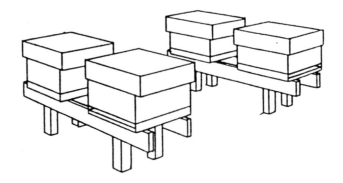

Fig 1   Apiary layout - Group of Four

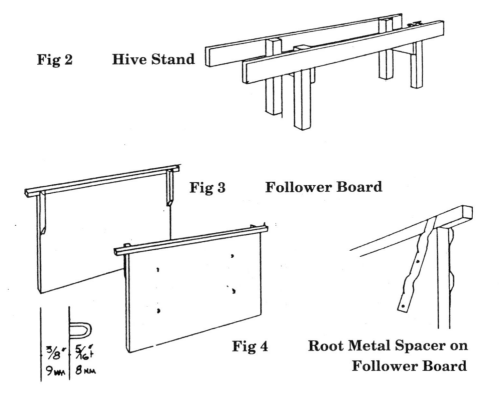

Fig 2   Hive Stand

Fig 3   Follower Board

3/8" 5/16"
9мм   8мм

**Fig 4   Root Metal Spacer on
Follower Board**

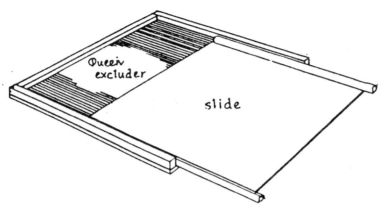

Fig 5   **Cloake's Queen Excluder/Slide**

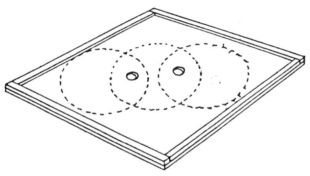

Fig 6   **Cover Board, with Feed Holes**

Fig 7   **Divided Hive for**
              4 x 2 Comb Nuclei

Entrances and feed holes

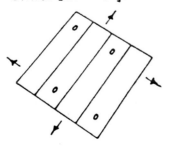

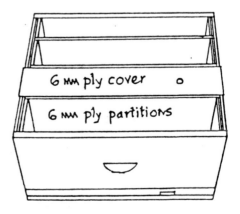

25

Miller method

A simple and suitable (and more certain) method for the production of up to about ten queens is that advocated by Dr. C.C. Miller, more than eighty years ago.[7] Here is what he wrote, in 1912: "Into an empty frame, two or three inches from each end, fasten a starter of foundation about two inches wide at the top and coming to a point within an inch or two or the bottom bar. Put it into the hive containing your best queen. To avoid having it filled with drone comb take out of the hive, either for a few days or permanently, all but two combs of brood and put your empty frame between these two. In week or so you will find this frame half filled with virgin comb containing young brood with an outer margin of eggs. Trim away with a sharp knife all the outer margin which contains eggs, except perhaps a very few eggs next to the youngest brood. Now put the prepared frame into the middle of a very strong colony from which the queen has been removed. In about ten days sealed queen cells are ready to be cut out and used wherever desired."

My own requirement is for about twenty queens a year, and I nearly always use the Miller method and rear two batches of ten. The provision of mating nuclei is often something of a problem. Miller's method requires at least two colonies, one a breeder colony for drawing the starter strips and providing eggs, and the other "a very strong colony from which the queen has been removed" to build queen cells on the trimmed new comb. The procedure I adopt makes use of at least three colonies, and makes provision for the necessary mating nuclei, as follows:

Select a colony strong in bees but preferably in a single box, with a queen that you wish to replace. Start feeding this colony a day or two after you have put into your selected breeder colony the frame with the foundation strips. (as in Fig 21).

When the comb built on the foundation is ready for the trimming operation, find and kill the queen in the colony you have selected for rearing the queen cells and have been feeding. Then transfer, from this colony, into a spare empty brood box, all combs containing brood, first shaking the bees from each comb back into the hive. Put the box containing these combs of brood, but no bees, on top of another colony, above a queen excluder. If the colony has supers above an excluder it can go above these. Bees from the colony will come up and care for the brood.

You can, if you wish, add combs of brood (without bees) from other colonies to fill the box. The combs of brood taken out of the breeder colony some days earlier could well have been similarly placed above a colony, and could now be added to complete the box.

The single box cell raising colony from which the brood combs were taken is now queenless and broodless, but has all its bees, and notably all its nurse bees, with no brood to feed. Cover this colony, and return to the breeder colony to remove and trim back to the tiniest grubs the new comb on which the queen cells are to be built, first brushing, not shaking, the bees off it.

When this comb is suitably trimmed, place it in the centre of the queenless colony. Carefully move the broodless combs remaining in this colony in to flank the new comb (the comb on each side preferably being a good pollen comb) and add combs (preferably containing some stores) to fill the box. The colony will have been fed for at least three days before this operation. Continue to feed.

A week later replace the cover board with newspaper (held down if need be by an excluder or by drawing pins) and on top of the newspaper put the box of brood that you put on top of another colony a week earlier, but this time with the bees covering the combs. Continue feeding.

Three or four days later, by which time the bees in the two boxes will be peaceably united, divide the colony into nuclei, preferably towards evening. As many nuclei can be made as there are combs of brood and queen cells, each queen cell being carefully cut from the prepared comb in the lower box and gently pressed (without damaging the cell) into a depression made near the top centre of each brood comb.

Diagram 1   Queen rearing by the Miller method

including provision of queen mating nuclei

On chosen day:

1 .Start feeding colony A selected for cell raising
2. From breeder colony B take all c br except 2, & g frame w f starters between these. Leave store c in B + all bs. Feed.
3. Put c br t from B (cleared of bs) in D above X on colony C. g dc to fill.

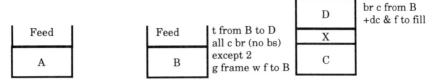

A week or so later (when c built on f starters is ready for trimming):
1. k q in A. t from A all c br (cleared of bs) & put these in new box E. put box E above D on C.   Cover A
2. Add c br (no bs) from other colonies to fill box E
3. Take from B the frame w f starters, brush bs off, trim c built to eg & ybr for q raising. put tr c in middle of A. g dc to fill A. Continue to feed A.

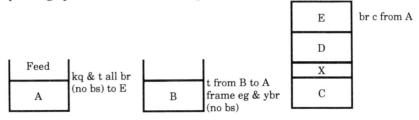

4. A week later, unite D & E to A through newspaper, continue feeding.

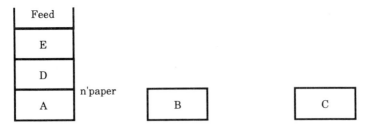

3 or 4 days later, towards evening , divide A, D & E into 3c nucs, each w 1 qc

28

If the nucs have to stay in the home apiary for the virgins to mate there, stuff the entrances good and tight and see that no bees chew their way out for a week. The nucs will come to no harm, and most of the bees will then stay put, even in the home apiary. One queen cell can be left in the colony with the bees on the comb and the flying bees, for the virgin to mate from the original site.

Alternatively the nuclei can be carried to a site beyond flight range of the home apiary, from which the bees will not fly back. Stuff the entrances with grass or moss so that it will take a day or two for the bees to chew their way out, and there will be little or no drifting between the nucs.

If provision has been made to ensure that there are plenty of drones flying of equally desirable parentage, so much the better.

Nucleus colonies from which virgin queens are to fly and mate should always be set on hive stands, to assist both the bees and the beekeeper. Two or three can be set on a standard hive stand, or they can be set individually on anything suitable. An upturned earthenware or concrete drain pipe makes a convenient stand. Assist the bees and queens to identify their nuc by facing entrances in different directions and colour marking them.

When the young queens are mated and laying and have capped brood the nucs can either be built up to form new colonies, which they will do surprisingly quickly with a little help, or used to requeen other colonies by de-queening and uniting, or even by uniting without de-queening since young queens seem always to kill older ones and not vice versa.

Even if you require only four queens this method is as good as any. If you wish you can reduce the number of queen cells being built in the cell building colony to the number you require, and thus ensure that they are well spaced apart and will be easy to remove and handle. You can also make up stronger nuclei. A four or five comb nucleus with two combs of brood and two or three combs of food is close to ideal, and builds up very quickly once the queen is mated.

Queen cell raising

It is sometimes convenient to raise queen cells for distribution to nuclei. e.g. to mininucs, or for direct introduction to colonies for requeening. The following two methods make use of a colony only temporarily, and allow the colony to continue normally as a honey production unit. I have used both successfully, and expect to do so again.

Harry Cloake's method.

This method, developed recently in New Zealand by Harry Cloake, and used in this country by Peter Kemble and others, is simple and neat, and has some advantages.[8]

A special board is required. It is very easy to make, as follows: Adapt a framed excluder (Herzog or B & J type) by lightly nailing to the upper side 19mm square strips along both sides, and the back, to match the sides and back of a brood box placed above, leaving the front open as an entrance above the excluder. Cut a piece of plastic laminate (Formica or similar) or sheet metal, as a loose fit to cover the excluder within the added frame, and fix two strips about 17mm square, or half round, to the upper side along the side edges, so that these strips project a little and serve both to keep the sheet flat and to provide a grip for sliding the sheet material in and out. That's all.

The excluder can later be put into normal use, without slide or added strips, merely by prising the strips off the excluder. Store the strips and slide for use when next required. (See Fig 5) An excluder framed to provide an entrance above it is also useful in other operations.

With this special division board it is possible, by inserting the slide, to divide the colony into two parts, queenright below and queenless above, and return it to a queenright state merely by removing the slide.

The method of operation is as follows: Choose as the cell raising colony a really good one with a good laying queen having the run of two deep boxes. Start to feed this colony about four weeks before cell raising is begun, and feed continuously until cell raising is completed or a natural honey flow occurs. Ample pollen must be available, or natural pollen must be fed.

A few days (up to a week) before cell raising is begun, turn the hive round so that the entrance is at the back, and rearrange the colony so that the bottom box has a full complement of combs, including three or four with brood and others with pollen and stores, and confine the queen to this box by the special division board from which the slide has been removed, i.e. by the adapted queen excluder. It provides a full width entrance above the excluder and this should face to the front, i.e. in the same direction as the entrance before re-arrangement. An upper box is placed on this board, and filled with three or four combs of brood, including young brood, and sufficient other combs of pollen and stores to make up to one frame less than capacity. This will allow space for the frame of cells. Add cover board, feeder and roof. Finally close the original entrance, now at the back. Then leave the colony undisturbed for at least two days. The bees will quickly adjust to using the entrance in the division board.

After two days, or a few more, on the day before cell raising is due to begin, the slide is slid into the division board, thus putting the bees in the top box into a queenless condition. Open the original entrance, now at the back. A less than full width entrance here is desirable, which a standard entrance block will provide. Bees flying from this back entrance will return to the hive by the front entrance and so add to the number of bees in the upper box.

Next day the cells to be started are introduced into the upper box. About twenty cell grafts in a frame can be introduced, or suitable queen raising material in another form, e.g. new comb with young grubs as in the Miller method. The following day the slide is removed and the rear entrance closed.

The colony is now queenright again, and the cells started on the cell grafts or other material can be left in the hive until ready to be introduced to queen mating nuclei. Except to check the brood in the upper box for rogue queen cells, which must be destroyed, it is not necessary to disturb the colony until the cells are ready to be removed.

Diagram 2   Cloake's method of queen cell raising

1. Start feeding over-wintered DD colony, AB, selected for cell raising, 3 weeks before start of operations.

2. Three weeks later, turn the hive around to face entrance to the back, confine q to the lower box A with Cloake board X, and close entrance, thus: Continue to feed.

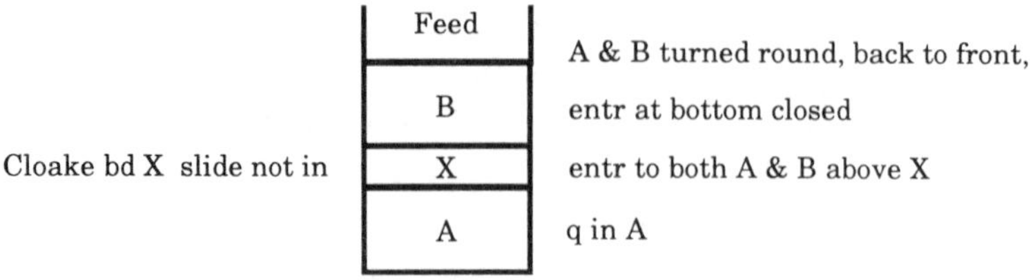

Feed	
B	A & B turned round, back to front,
	entr at bottom closed
Cloake bd X slide not in X	entr to both A & B above X
A	q in A

3. Two or three days later, insert slide, & open bottom entrance

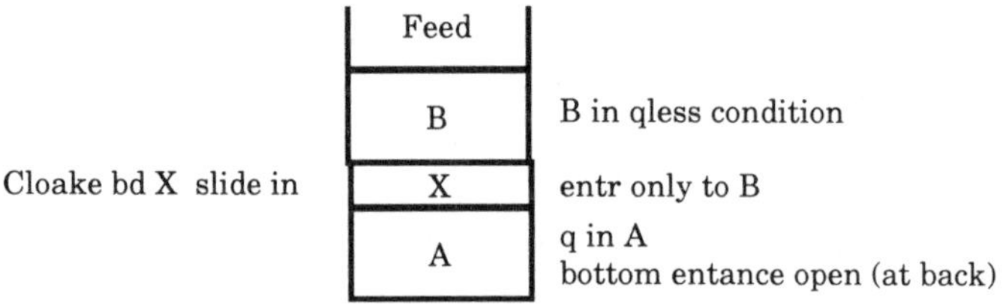

Feed	
B	B in qless condition
Cloake bd X slide in X	entr only to B
A	q in A
	bottom entance open (at back)

4. Next day, g cell grafts or Miller trimmed c to B

5. The following day, remove slide and close bottom entrance

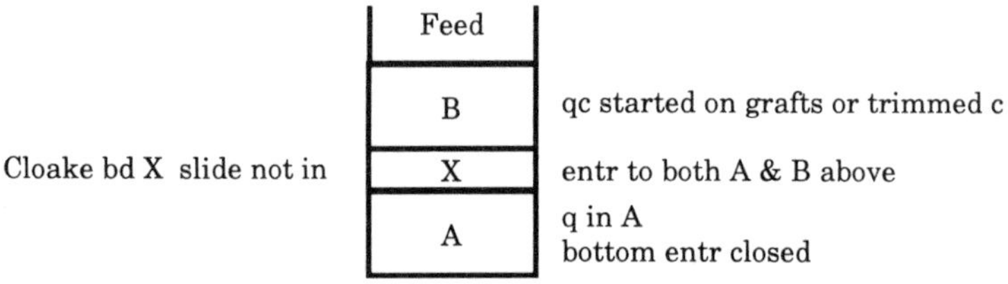

Feed	
B	qc started on grafts or trimmed c
Cloake bd X slide not in X	entr to both A & B above
A	q in A
	bottom entr closed

6. When ready for use in q mating nucs remove qc, remove Cloake X board, turn hive round and open bottom entrance. Colony now DD, &, as originally, q has access to both A & B.

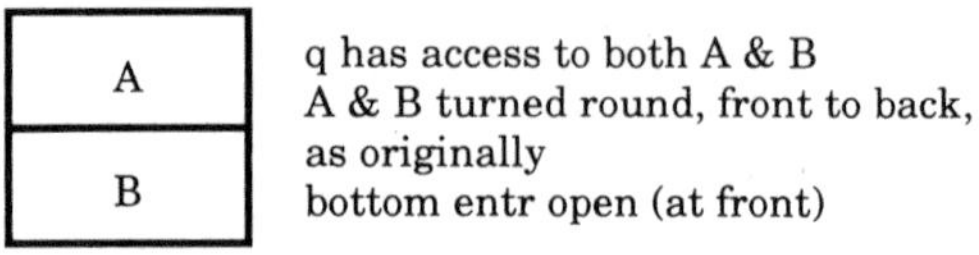

A	q has access to both A & B
	A & B turned round, front to back,
B	as originally
	bottom entr open (at front)

If desired, the colony can be used to raise a second batch of cells, either when the first batch is removed, or as soon as the first cells are sealed. The procedure is as before, that is, place the slide in position the day before grafts are introduced and remove it the day after introduction.

When cell raising is completed, the division board is removed. and the colony becomes a normal honey production unit of undiminished strength.

Vince Cook's method.

A method with a similar objective, namely to permit the cell raising colony to continue as a honey production unit, has been developed by Vince Cook, as follows:[9]

Choose a good colony in a single brood box, say with brood in 8 combs, move the colony to one side and put a new floor and brood box in its place. From the colony take a comb of brood with the queen and adhering bees, together with another comb with brood, pollen and bees, into the centre of the new box, and fill with empty brood combs. Add an excluder. Put two empty combs into the old box to replace those taken, and put the old box above the excluder. Add a feeder, and feed. Leave undisturbed for a week.

Then, at 10 a.m., replace the upper box on the floor, remove one comb without brood from this box and leave space for a comb in the centre of the box. Add a cover board and feeder (full) and a box to enclose the feeder, and above that a split board with a small front entrance. On this put the box with the queen. At 2 p.m., graft into cell cups (about 24 on 2 bars) and put the frame with the grafts in the space in the bottom box. Alternatively, put in a trimmed new comb with tiny grubs made available at this time by following the Miller method. Rebuild as at 10 a.m.

Next day, at 10 a.m., re-arrange the hive, putting the box with the queen on the floor, excluder above, box with grafts or trimmed comb above the excluder, then the feeder and its enclosing box, cover board and roof.

Ten days after grafting or insertion of the prepared comb finished queen cells are ready for distribution to nuclei, and the colony can continue as a honey production unit with undiminished strength.

Diagram 3   Vince Cook's method of queen cell raising

1. Move selected colony in single box A to one side and put empty box B in place. Take c w q & adher bs, + c w br, pollen, & bs, from A & put both c in B. g dc to fill B. Add X & put A above. g 2dc to fill A. Feed.

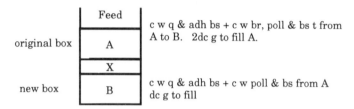

	Feed	
original box	A	c w q & adh bs + c w br, poll & bs t from A to B. 2dc g to fill A.
	X	
new box	B	c w q & adh bs + c w poll & bs from A dc g to fill

2. One week later, at 10 a.m., set aside feeder, box A, and X. Set aside box B and set box A in its place. Take 1 c w no br from A , and rebuild as below:

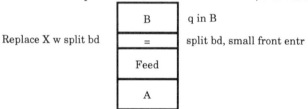

	B	q in B
Replace X w split bd	=	split bd, small front entr
	Feed	
	A	

and, at 2 p,m. that same day, g cell grafts or Miller trimmed c to A.

3. Next day, at 10 a.m., re-arrange, exchanging A & B, as below:

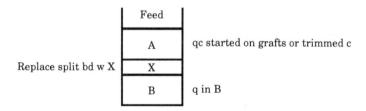

	Feed	
	A	qc started on grafts or trimmed c
Replace split bd w X	X	
	B	q in B

Ten days later, take qc from A for use in q mating nuclei.

Queen Removal

An even simpler method of raising queens from a selected breeder which I have used from time to time myself and which I know to be the preferred method of at least one commercial beekeeper, is merely to remove the queen during a nectar flow. Thus made queenless, the colony will build queen cells on larvae of its own choosing and nucs made up nine or ten days later are each given one of these cells.

More explicitly, proceed as follows: Select a strong colony headed by a queen which you are happy to use as a breeder, and operate during a nectar flow; that from oil seed rape for example. Remove this queen, taking her into a nuc box with two combs of brood and a comb of honey and pollen and the adhering bees. She will thus continue to lay, and can be used for further production of queens. Deprived of its queen, the colony will begin to build queen cells. Most such colonies, in my experience, can be relied upon to provide ten or twelve good queen cells.

Nine or ten days later, by which time the queen cells will all be capped but no virgin will be ready to emerge, make up the requisite number of nucs from this and other colonies and give each one a queen cell. Make up three-comb or five-comb nucs, each with two combs of sealed brood and one or more combs of honey and pollen. Take them to another apiary for the queens to mate, far enough away so that bees don't fly back. If the nucs have to stay in the apiary in which they were made, stuff the entrances tight with grass and ensure that no bees fly from them for three or four days (but not much more). They'll come to no harm from a short confinement.

Making and using nuclei

One comb of emerging brood, with a queen cell and one or two combs of food, with ample bees to cover, will make an satisfactory nucleus colony. Convenient accommodation for such nucs can be provided by a single walled brood box (National, Smith or other) with bee tight divisions inserted to make three 3-comb or four 2-comb nucs, and a floorboard made from a standard inner cover or from a sheet of plywood with a ½ inch deep rim around the upper side. Pieces cut out of the rim provide a small entrance to each nuc on the three or four sides of the box. (See illustration Fig 7)

Such nucs can be placed in any convenient position in the apiary (with the entrances stuffed tight) and provided with a roof. If the box of nucs is put on top of a colony the need for another roof is removed. Three 3-comb nucs in a box can have entrances to the back and to each side.

Unless there is reason to do otherwise, do not inspect newly made up nucs until the queen can be expected to be mated and laying. In favourable weather a small nuc can be opened without smoke, or with almost none, so that the queen can easily be found.

When bees are flying freely from the nucs an empty shallow super can accommodate 3 or 4 small lever lid tin or glass jar feeders (Golden Syrup tins or jars are just right) and the nucs can thus be conveniently fed as required. I drill about 20 holes in the lids with a 1/16 inch drill, in preference to using a nail to make the holes.

Once the young queens are mated and laying it will not be long before a 2 or 3 comb nuc will become crowded. One nuc can be built up in the box, when the divisions have been removed, (and united to the colony below, if desired) and the others used elsewhere.

If one of the queens fails to mate, two of these small nucs in a divided box can safely be united by removing the division board between them and replacing it with a follower board for a few days, and then moving the follower board to its usual place at the side.

All my nucs are made with standard brood combs. This suits my operations better than would half frames or mininucs, and also, by so doing, I almost never need introduce a caged queen to a colony, other than a bought queen. I unite the nucleus to the colony. For me, therefore, the choice lies between nucleus boxes, to hold up to five combs, and divided brood boxes, to hold three or four nucs in each. I have both.

A four or five comb nuc is costly to stock and maintain, but will hold a queen for several weeks, and is large enough to be transferred to a brood box and built up as a colony if so desired. The adaptation and use of a brood box to take three or four nucs requires a special bottom board, division boards and covers, but will take standard boxes to enclose feeders, and a standard roof; and the smaller nucs are convenient in use for requeening colonies. I have ten nuc boxes to hold five standard brood combs, and also have the necessary floors, division boards, covers, etc., to adapt a brood box to take four 2-comb nucs or three 3-comb nucs.

My 5-comb nuc boxes are usually stocked with 4-comb nucs and two follower boards. A division board feeder can then replace the follower boards if required, and a fifth comb can be given later. If queen rearing is undertaken at apple blossom

36

time and nucs of this size given some assistance and feeding they will make excellent colonies for the heather. Well fed, 5-comb nucs can safely be overwintered.

I commonly assist nucs to build up, once the young queen is mated and laying, by giving them young bees from a colony that can spare them. Take a comb of brood from the brood chamber, make quite sure that the queen is not on the comb (preferably by seeing that she is elsewhere), and shake it just sufficiently to dislodge the flying bees before carrying the comb to the nucleus and shaking the young bees (which always cling more tightly than do flying bees) into it. Repeat the operation with a second comb, if you think fit. Return the shaken combs to the brood nest where they came from. Repeat the operation a week or so later.

If a sloping board can conveniently be set against the entrance of the nuc, the preliminary shake to free the comb of the older bees is unnecessary, as they will fly back to their colony, leaving the young bees to crawl into the nucleus.

A good method of requeening by uniting a nuc to a colony is to remove a sufficient number of combs from the brood chamber to permit the whole nucleus colony to be put in their place. Carried out on a day when bees are flying freely, an hour or so after the queen in the colony has been removed, the method is quite reliable, in my experience. With Hoffman frames, grasp and carry three or four combs as a unit, without separating the combs.

Pulled virgins

I ought to say something about pulled virgins, which H.J.Wadey first introduced me to about 1946. He devotes a short chapter to their description and use in his delightful (and I suspect much underrated) little book "The Behaviour of Bees - and of Beekeepers".[10] As he says, a pulled virgin is a young queen that has emerged from her cell after it has been removed (hence "pulled") from the hive.

Carry a few well aired empty matchboxes in the bee box, or in your pocket, and you can, if you wish, make use of virgins on the point of emergence from the cell by putting one such cell in each matchbox and allowing the queen to emerge there. Provided a colony has no adult queen, a pulled virgin allowed to run into the colony from the matchbox, having met no other bee since emergence from the cell, will almost always be accepted in the colony.[11]

I rarely use one myself, but I have done, and it is helpful to know of the possibility. Certainly there are occasions when one finds virgin queens emerging from queen cells during an inspection, e.g. when the colony has swarmed, and has a well aired match box handy.

Wadey says that if a queen is present in the colony the virgin will be killed. It is not always so, in my experience, and I have twice known a bad tempered colony, in which the queen could not be found, successfully re-queened by the simple expedient of running a pulled virgin into the entrance. It is worth trying, as nothing would be lost if it did not succeed.

Pulled virgins can be kept in the matchboxes without harm for several hours in a warm place such as one's trousers pocket, and Wadey says they can be kept for up to a week if kept in a warm, dark, place and given a good drop of diluted honey daily. Ernie Pope was always popping "ripe" queen cells into matchboxes and giving pulled virgins to local beekeepers in trouble, and many have re-queened their colonies with his very good strain of bee by that means. I know he has sometimes kept such virgins for several days.

Bought queens

Routinely, I re-queen colonies at the end of their queen's second year. I seldom buy queens for this purpose; I rear them myself from a selected breeder queen. Breeder queens should be kept in small single box colonies, kept small by removing brood from time to time. Successful introduction of a bought one is vital, and the best way to achieve it is to introduce her to young bees in a small nuc.

From a suitable colony take a comb about half filled with brood, including young brood, and preferably with brood in all stages, and take two or three other combs containing pollen and honey. Shake the bees from all these combs back into the colony. Replace the combs taken with empty combs. Put the brood and food combs cleared of bees back on the colony or on another colony in a box above a queen excluder. Young bees will soon come up to cover the brood.

38

An hour or two later, preferably towards evening, carefully take the combs that you put above the excluder with the adhering bees and put them in a 5-comb nuc box, with a dummy board on one or both sides. Stuff the entrance with grass. Give the bought queen to this nuc in a Butler cage, without attendant bees, hung between the comb with brood and its neighbour. Leave the nuc alone for a week, by which time the bees should be flying from it and the cage can quietly be removed.

Re-queening colonies with bought queens in late summer or autumn is also best done by introducing the queen without attendant bees in a Butler cage, as follows:

1.   Find and kill the old queen (or take her into a queen cage with some attendant bees if she is to be kept).

2.   Leave a gap between two brood combs (preferably adjacent to the comb from which the queen is taken) sufficient to permit the introduction of a Butler cage. The removal of the follower board makes this easy. Replace the cover board. (Repeat these operations on a second colony if more than one queen is to be replaced).

3.   In the car, with all windows closed, release the new queen from the travelling cage, run her into a Butler cage and confine her there, unattended, with a small piece of tissue paper held in place with a rubber band. The attendant bees commonly leave the travelling cage before the queen, and the queen will walk into the Butler cage with a little encouragement. If she flies, she can be picked off the car window and run into the cage.

4.   Returning to the colony, remove the cover board and suspend the cage in the gap left between the brood combs, using the minimum of smoke. Replace the cover board and roof.

5.   Leave the colony alone for a week, or a little more, then quietly remove the cage, close the brood combs together, and replace the dummy board, cover board and roof.

The interval between the removal of the old queen and the introduction of the new one is not critical, provided it is not very long, say not more than an hour. I have more than once found and killed four queens, then returned to the car and caged four new ones, and then introduced these four, all successfully, the interval varying between, say, 50 minutes for the first and

10 minutes for the last colony. In my experience, queen introduction in this manner in late July or early August is very rarely unsuccessful.

The Butler cage is well known and widely used. As I use it, the cage is suspended by frame nails (see Fig 19) between two combs, for which the removal of the follower board leaves ample room. I sometimes use the Worth cage (Fig 18) which is similarly suspended, and which I use in the same manner, and I have a few Manley cages (Fig 20) that I use also. All three kinds of cages are intended for the introduction of queens unaccompanied, which I (and many others) think best.[12]

Introduction of a bought queen in the mailing cage, without removing the attendant bees, is usually successful when the introduction is to a small nuc with a preponderance of young bees. Put the mailing cage vertically between the combs, screen exposed, candy at the top, and allow the bees in the colony access to the candy.

Queen rearing, and the associated endeavour to keep and maintain the best bees you know, is a fascinating subject, and very well worthwhile. Manley's assertion "that the key to getting honey under adverse conditions is to breed your own queens" is founded on much experience.[13]

New Zealand Queens

For the past twelve years, since queens from New Zealand became available, I have bought a few of these newly mated queens in late April each year. At first I had doubts about the suitability of New Zealand bees for conditions here, and bought only two or three queens in each of the first years. But experience proved their worth, and more recently I have had six or eight of these queens each spring. Their arrival coincides with the first show of blossom on oil seed rape. I introduce these queens to eight or ten comb nucs made up from the strongest colonies that might otherwise swarm at the rape. It is an effective method of swarm control, and a simple way of replacing winter losses. Such nucs build up rapidly on the abundant pollen and nectar from the rape, and soon become productive colonies that do well, winter well, and perform well in the following year.

To each of the brood boxes I require for the nucs, I clip a floor board, and provide a screen, inner cover, entrance block and roof. In each one I put a frame feeder, a dummy board, and a full complement of good drawn combs (nine in a Smith box). When the queens arrive I load up these boxes, and a spare empty one, and take them and the queens in their cages, to a selected out-apiary. There I unload the spare empty box and two or three of the nucs, open one of them and take out all the frames, put these in the empty box. I leave only the frame feeder and the dummy board. Then I move to the first hive from which I intend to take a nuc. What I will take, if I can, is three good combs of brood, and four, five or six combs of honey and pollen, all with their adhering bees, but I must first find the queen in the colony I am taking them from, and make sure the queen stays behind.

I put the roof upside down on the ground nearby, prise the upper box loose from the bottom box without disturbing the cover board, lift off the upper box and set it slightly skew-wise on the upturned roof. I then take out the dummy board, and proceed to examine carefully the combs in the bottom box, looking for the queen. If the queen is not found in the bottom box I must search for her in the upper box. The aim, when the operation is completed, is to have made up a nuc with three good combs of brood and four or more combs of honey and pollen, all with their adhering bees but with no queen; and to leave the colony from which the nuc has been taken with the queen and the remaining brood combs in the bottom box, and mainly empty combs, taken from those brought with me, in the upper box. The nuc is screened and the entrance closed. I then move to the next colony and repeat the operation. Moved away and given one of the new young queens, each nuc will build up rapidly and soon become a productive colony.

On the new site I transfer each queen - in the car, with the windows closed - from the travelling cage to a Butler cage, and put the queen thus, unaccompanied, between two of the brood combs in each nuc. The colonies from which the nucs have been taken will have lost very few flying bees, will expand into the nearly empty upper box and will be unlikely to swarm for some time.

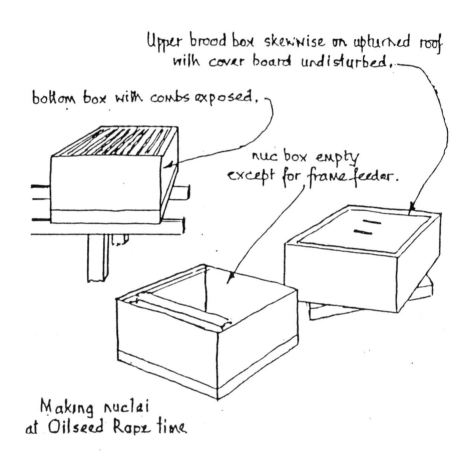

Upper brood box skewwise on upturned roof with cover board undisturbed.

bottom box with combs exposed.

nuc box empty except for frame feeder.

Making nuclei
at Oilseed Rape time

Sketch of Making Nuclei at Oilseed Rape Time

Myself in Colin Weightman's home apiary in 1964
Note smoker held between knees

My partner and me in one of my apiaries in Essex in 1993

43

Fred Richards in a Buckfast Abbey out-apiary in 1961
Note the arrangement of the hives, and the wind shelter

Buckfast Abbey hives on Dartmoor for the heather
Note wind shelter

44

14x12 frames and boxes newly assembled

**Brood comb top bars, untouched, showing brace comb
only where narrow top bars in use**
Note dummy board in position

Brood box, divided to take four 2-comb nucs

Manley shallow frame, with extra thin foundation for cut comb

46

3   Handling bees

Are not these matters of sufficient importance,
and difficulty to require learning by experience
and careful inquiry, just as much as any other art?

Florence Nightingale, Letters.

Handling bees

The knack of handling a colony of bees can be learnt;
indeed it must be learnt if one is to become and remain a
beekeeper. The best way of learning is to work with and assist a
competent beekeeper who can also instruct and who has well
behaved bees; and the next best is perhaps to buy a nucleus
colony of really docile bees (and there are some) and teach
yourself.

But some bees are impossible. In 1947 a farmer friend at
Nutley, near Maresfield, in Sussex, begged me to take his six
colonies, as he considered them quite unmanageable, and so
they proved to be. The bees had stung him and his wife, his
children, his workmen, and his livestock, and he daren't go near
them. They were well sited in a clearing on the edge of a wood
in almost new Modified National hives and fully half a mile from
the farmhouse and farm buildings and from the public road, a
little used lane. Even in a good nectar flow one could do little
more than add a new super before being attacked, no matter
how much smoke one used. I never removed veil or gloves until
I and my Land Rover was the best part of a mile away, as the
bees would chase in fury. I managed, eventually, to requeen

these colonies by moving them away, one at a time, to lose their flying bees and allow me to find, and destroy, the queen.

Fortunately, bees are very rarely vicious to anything approaching this degree, but I didn't then, and don't now, like bees that "follow", even if they don't sting.[1] Do learn to handle bees with bees of a docile strain, preferably by working with a competent and experienced beekeeper, and don't keep bad tempered bees.

The strain of bee is all important. With good bees colony inspections present little difficulty and management is straightforward; with a poor strain, and particularly with a bad tempered one, the whole business gets neglected. All too many inspections are curtailed or not undertaken because the beekeeper is afraid of his bees, and the situation gets worse as propolis, brace comb and burr comb accumulate and the supposedly movable frames become anything but movable. So do start with docile bees, try hard to keep none but good tempered bees that will go about their work without bothering the beekeeper or his neighbours, and try to learn to handle bees competently and with minimum disturbance.

It is not merely the control of the bee colony that has to be learnt, but also the art of combining speed of manipulation with gentleness and thoroughness, and that requires practice, trained observation, and concentration. If the queen is being sought, then concentrate on looking for her and on nothing else.

Don't open a colony without a good and sufficient reason, have clearly in mind what your purpose in opening it is, and keep that purpose in mind.

Don't operate without a veil, or even think of doing so. "No sensible person will open hives unveiled" wrote Manley (in Bee Craft) So be sensible, and always wear a veil.

Wear gloves too, if you must, (but not always), and certainly do so if you intend to handle a queen - when you should take them off, so that you can pick her up with clean fingers.

Inspections

Light the smoker well in advance of use, ensure that it is going well, and have spare fuel to hand for replenishment when you need it. Work from the side or the back of the hive. Do not obstruct the normal flight path of the bees. Avoid sudden, jerky, or rough movements; make them smooth and deliberate. Except with intent, e.g. to shake bees from a comb, do not handle hives or frames roughly.

Keep all boxes covered. Spare cover boards, and cloths a little larger than the surface area of the hive, with a rod in a pocket along two edges to weight the cloth down, are useful for this purpose.

Give smoke at the entrance, and then beneath the inner cover or between the boxes, a minute or two before removing the cover or a box. Prise off the inner cover, or prise boxes apart, with care (freeing them at more than one place) and puff smoke across the frame tops as you do so.

If boxes are heavy, or likely to be, it is useful to have something on which to stand them at a convenient height. Putting such boxes down and lifting them up is heavy work; it is much easier to move them to a position near to their original level. An empty box or two on the upturned roof, or two or three roofs piled up, may serve. Some beekeepers make a portable stand on which to put the boxes during manipulations.

Keep an eye on the boxes set aside (which should be covered) and give a puff of smoke to them if need be. Puff smoke occasionally (as may be necessary) across the frame tops of the box on which you are working.

Unless conditions are such as to make it unwise to delay (bees cross, robbing likely, etc.) scrape burr comb from the frames before returning them to the box, thus keeping the frames more readily removable. Put the scrapings in a tin and cover. Do not allow scraps of comb to remain on the ground and incite robbing.

Make a note of what you find, and of what you do. Learn to work quickly, so that hives are open as briefly as possible. Open hives only when necessary, and with good reason. When hives get tall take a strong box around with you to stand on; a spare deep roof or an apple box is what I use.

Inspection routine

It helps greatly to learn a routine of inspection, so that it becomes habitual. Proceed as follows: If the colony has an excluder and supers, remove them first, in one piece if possible, and set them aside. Then remove the upper brood box (assuming there are two) and set that aside and cover. Then examine the combs in the bottom box; replace the upper box and examine the combs in that box; and finally replace excluder and super(s). Then see what stage work in the supers has reached.

It may be helpful to some to describe these operations in more detail. First of all, get the smoker going really well. Then open your record book at the page appropriate to the colony that you intend to inspect, remind yourself of its condition at the last inspection and of what you expect to find this time. See that you have conveniently to hand whatever you think you may need.

Give a puff or two of smoke at the entrance, but thereafter smoke from the top, across the frame tops, as may be necessary, using only just sufficient smoke to control the colony but not too little to keep it under control. Only experience can teach the art of smoking. Remove the roof, and set it aside, upside down, in a convenient place to receive any boxes subsequently removed from the colony. If there are two colonies on each stand, then give a puff or two of smoke at the entrance of the second colony and remove that colony's roof also to serve as a second stand for boxes, should the need arise, or as a cover.

With the smoker going well, insert the hive tool below the excluder near one corner at the back and prise up the excluder and supers a little, give a puff of smoke, and allow the crack to close again. Insert the hive tool below the excluder near the other corner at the back, prise up a little, give a puff of smoke and once more allow the crack to close again. If need be, prise up once more, give more smoke, and allow to close. Then set the smoker down and lift excluder and super(s) off as one piece and set aside crosswise on the upturned roof.

Smoke under the excluder and across the frame tops of the brood chamber. Prise the two brood boxes apart with the hive tool (again with smoke) and lift and set that box aside. It may conveniently be put on top of the super(s). Cover it with a spare cover or with the second upturned roof. Then proceed to inspect the combs in the lower brood box.

Remove the dummy board and prise the first frame away from its neighbour. Lift it straight up and examine it. Replace it in the hive close to the hive wall with the inner face inwards, i.e. facing the same way as it was before you took it out. The removal of the dummy board will have left ample space. Repeat for the next frame, replacing it against the first, and again facing the same way as it was. Repeat for all other frames.

Then replace the dummy board and exert pressure with the hive tool to move all the frames across to make firm contact with each other and with the dummy board and with the side wall of the hive. Exert as much pressure as need be to maintain the desired frame spacing.

Now replace the upper brood box on top of that which you have just examined, and repeat the inspection process. Finally, replace excluder and super(s), and make such inspection of the supers as you think necessary. Replace the inner cover and lockslides but not the roof.

Now record, in your record book, what you consider relevant. I always record first, whether I saw the queen, and if I did I write "sq". I then record the amount of brood and stores and whether there were queen cells, e.g. "9br d br in 4 up st OK no qc." And so on.

Proceed similarly with the second, third and fourth colonies, replacing roofs as convenient, and move on to the next group of four, and repeat.

I would emphasise two details that many neglect to observe but I consider desirable. First, to ensure that combs are returned to the brood box in the same facing direction and in the same order in which they were. Change, in either respect, should be deliberate and with reason, and not inadvertent.[2] Second, to treat the excluder and lower super (at least) as one piece. Handle the excluder separately only when necessary, both to minimise operations and to avoid unnecessary breakage of brace comb.

The inspection process should serve two related purposes. One is to observe, evaluate and record the combs as combs, i.e. their suitability for continued use (completeness, flatness, drone comb, etc.) and as part of the brood nest, (honey, pollen, amount and stage of brood, queen cells, abnormalities, etc.). The other is to observe, evaluate and record the state of the brood nest and the colony as a whole, i.e. presence of queen, amount of brood, stores, etc., presence or absence of queen cells, health of colony and honey in supers. What one discovers will, of course, largely determine what action is called for, either on that occasion or later. As Coverdale says, perceptive observation is a most valuable skill.[3] Like other skills, it has to be learnt and developed.

The first inspection in spring should be a really thorough inspection. Every comb in the brood chamber should be examined and the state both of the combs and the colony carefully assessed and noted. Inspections thereafter can be as brief or as thorough as circumstances suggest or require.

It may happen that bees become so alerted (both in the colony on which you are working and in others) that continued operation is no longer possible. If a colony gets out of control, reassemble the hive as quickly as possible, remove all combs or boxes to which bees could get access, and leave the apiary. Do not attempt to open that hive or other hives in the apiary that day.

It may also happen that a colony is difficult to handle, or to keep under control, stings passers by, and should be eliminated or re-queened. If it can be moved ten or twelve metres from its stand the necessary steps can be taken more easily. Proceed along the following lines, preferably on a day when bees are flying well:

If the hive has supers, first remove them and set them aside with a screen beneath and a cover on top. Then move the hive from its stand, perhaps near to a weaker colony. On the now empty stand put a box with frames and foundation, and in this box put one comb of honey and pollen and a little brood (preferably from another colony) and replace the supers on top. The flying bees return to a queenless and almost broodless box, and will be much less difficult to handle. Next day, or the day after, if bees have flown well, all the flying bees will have returned to the old stand, leaving only young bees (that are far less likely to sting) and the colony can be inspected, the queen found, removed and replaced.[4] If a site is available, or can be found, beyond flight range, the colony can then be moved again, to that site, losing more bees to a nearby colony.

Stings

Most beekeepers develop a degree of immunity to the effect of bee stings, but the pain which a sting inflicts is not, I

52

think, diminished at all. I, and other beekeepers of long experience find a bee sting just as painful today as ever it was, although the effect of one, or of several, may be scarcely noticeable.[5]

A beekeeper needs to develop confidence in his ability to handle colonies of bees and to carry out the various necessary and desirable operations. Good tempered bees are essential if confidence, and competence, are to be acquired. A competent beekeeper is not often stung. Never allow a colony to get out of control.

Gloves

It is certainly not necessary, and I do not consider it wise, always to wear gloves when handling bees. Impenetrable armour can lead to insensitive handling and poor judgement, and to the toleration of bad tempered bees. But regular inspections, particularly in out-apiaries, have often to be undertaken in much less than ideal conditions, when gloves are a near necessity.

Good gloves, thin and pliable, but tough, can, with practice, positively aid sensitive handling rather than otherwise. I wear gloves as much to keep my fingers clean as for any other reason, but the confidence of beginners is greatly assisted if gloves are worn, and I consider it unhelpful to advise against their use. With experience the beginner will come to know when to wear them and when not.

If queens are to be handled for clipping, or any other purpose, then always wear gloves, and remove them only when handling the queens. Without gloves the inevitable propolis on the fingers can make handling queens quite difficult and potentially damaging.[7]

Removing supers by shaking

It is sometimes necessary, or best, to remove supers by shaking bees from the combs instead of by using a clearer board, and it is as well to know how to do it.

Around tea-time on a cool day, when bees will not be flying much, is a good time to operate. Smoke from the top, remove and set aside the top super and replace the cover board of the colony. Have ready a spare empty super and two or more cover boards, and stand the super on one of these boards so that bees cannot gain access from the bottom. Then remove a comb and shake (and if need be brush) the adhering bees into an upturned roof. Put this comb into the empty super. Repeat the operation until all the combs have been cleared and transferred, and put a cover board on top to make the super bee tight. Use the emptied super for a repeat operation with the second super and again cover.

Bees shaken in this way - a sudden severe jerk to dislodge them and throw them into the roof - don't fly much on a cool evening, and after shaking two supers there will be two or three pounds of bees in the roof. They can be shaken - again with a sudden severe jerk - in front of the colony from which they came. Carry the cleared supers into a beeproof place for extracting, or stack them on the truck or trailer so that they are beeproof, then repeat the whole operation with the next colony. Do not continue if your handling of supers from the first colony has caused other colonies to become interested. Robbing is a lot easier to prevent than to stop. Better leave the second colony until next evening, or use a clearer board.

Swarms

It is necessary to acquire some competence in taking swarms, even if the number of one's own colonies that swarm is close to none. The removal of an unwanted swarm is a service to the community which few beekeepers can, or would wish, to avoid. Usually, the job presents little difficulty, but sometimes it borders on the impossible. No swarm is worth a broken neck.

Essentially, a swarm is taken in one of two ways, either by shaking the swarm cluster into a box or other receptacle, sometimes after removing the branch or whatever it is clustered on and carrying it away, or by encouraging the cluster to move to a more convenient place, or on to a comb or combs, from which it can be more readily taken.

It is useful to have a suitable receptacle ready to hand, particularly one in which the swarm can be transported without difficulty, whether as a shaken cluster or on a comb or combs, such as a travelling box, part screened, with a clip-on removable lid. A 5-comb nucleus box, part screened, would serve, or a screened broodbox.

A single comb in which bees have been bred, or preferably a comb with a little brood in it, placed close to, and preferably just above, the cluster, is attractive to the bees. They will usually move on to the comb of their own accord, quite quickly, and can be encouraged to do so with a little smoke. In due course the comb with its adhering bees can be put in the travelling box, in the almost certain knowledge that the queen will be on the comb. The remaining bees will join her there, and the whole lot can later be carried away.

With a little ingenuity and a length of cord, a comb with a little brood in it can be manoeuvred into, and later out of, the most unlikely places. For example, with a weight at one end the cord can be thrown over a branch, the weight then replaced by the comb of brood, pulled up to attract the swarm, and lowered with the swarm clustering on it.

I have assisted some very ingenious swarm takers, in my time, and have exercised ingenuity to take a few myself.

4   Hives and equipment

The most important unit of beekeeping equipment is the hive.
The hive is merely a tool to assist the beekeeper,
not the bees. Some are better tools than others.

R.O.B.Manley, "Honey Farming" (1946)
and in Bee Craft (1968)

The choice of hive

I started my beekeeping before the war with Langstroth
hives, but I have had Smith hives exclusively since 1949. I still
have in use some of the brood boxes and most of the frames that
I bought from Yorkshire Apiaries and Mountain Grey Apiaries
in 1947. Some of the brood boxes, made of Parana pine, shrunk
overmuch, and were cut down for supers and overall feeders.
The frames have been stripped and re-fitted with wax
foundation countless times, and are good for another forty years.
The boxes and frames bought since those early years will last
just as long. With reasonable care in use, beehives and frames
will last a lifetime, and perhaps two lifetimes, so that the choice
of hive and frame is of the first importance.

If we lived in the USA or Canada, in Australia or New
Zealand, or even in Mexico, the problem would scarcely arise, as
the Langstroth hive and the Hoffman frame is so nearly
universal. But we don't, and the choice open to us is confusing,
to say the least, and not at all easy. Frankly, I doubt if the
choice is often logically considered and made. More usually the
new beekeeper acquires his bees on combs and in a hive or

nucleus box and goes on from there, and in the UK the likelihood is that they will be 14 x 8½ frames with long lugs in a bottom bee space box.

No doubt it is largely because beehives and frames do last so long in use that so many British beekeepers never escape from the stranglehold of the WBC hive. National and Modified National hives take the WBC frame and its metal ends and even adopt the undesirable bottom bee space. Indeed the Modified National hive is essentially a WBC inner box made of heavier material and thereby able to dispense with the outer case. For those who desire to use the 14x8½ frame size (and there are good arguments for doing so) only the Smith hive embodies the undeniably evident and very real advantages of top bee space and truly single walled hives (which on that account requires a short lugged frame), and it has largely been the logical Scots and their neighbours the Geordies who have adopted it.

In the UK as a whole there is little doubt that Modified National and National hives greatly outnumber all other types.[1] Because beehives remain serviceable for such a long time, this situation is likely to continue for many years, perhaps indefinitely. For those who may wish to sell colonies of bees, or nuclei, and those who think it possible that they may give up beekeeping after a short time and then wish to sell, the arguments for using Modified National hives are strong. The hive is in such widespread use that there is likely to be a ready sale for colonies in such hives, and for good secondhand equipment to suit. It is far less so for other makes and types.

Assuming that decision, then at least the frames can be Hoffman frames, with 1 1/16 inch wide top bars and bottom bars of full (7/8 inch) width (as they should be) and remain "standard". To do so does no more than adopt the frame that is now widely in use. At least, also, metal ends on other frames can be replaced by Hoffman clips.[2] Change to a top bee space (which I should personally consider desirable) departs from "standard" and would reduce opportunities for sale. The widespread adoption of Hoffman frames and clips, to an extent where it becomes the "norm", seems certain, but the bottom bee space seems likely to remain a feature of most National hives for as long as they are in use.

Frame size

In fact, a well based decision as to which frame size to adopt is not easy and straightforward. It hinges mainly on whether the requirement is for a single box large enough at all times to serve as the brood chamber of a prosperous colony, or for the flexibility that a brood chamber of more than one box can provide. Probably only Modified Dadant and Langstroth Jumbo hives meet the first requirement.[3] The Modified Commercial, with a 16x10 frame, may do so for a bee that is only moderately prolific. It is a good hive, used by more than a few beekeepers. The frame is a good shape and of adequate depth, and the hive can hold twelve frames with 1 3/8 inch spacing. But do have it with a top bee space, as you can if you ask.

The Langstroth frame has almost exactly the same surface area as the 16x10 but is a less satisfactory shape and of inadequate depth for single box working, and the hive takes only 10 frames. Universally it is used as a multi brood box hive, and two Langstroth boxes provide an unnecessarily large brood chamber for use in this country, or so I think. But I can well recall H.J. Wadey's double Langstroth colonies in Sussex, and the very good crops that he had from them, and it could be a good choice.[4] The Langstroth has one enormous advantage, namely that it is so widely used that a large part of the world is open as a source of supply, and open to our own manufacturers as an export market, so that we could expect the cost and availability of equipment to be advantageous.[5]

It is as well to know, what many seem to be unaware of, that Langstroth frames and boxes, standardised in all other respects, are offered in several depths, and not merely deep and shallow. The six common frame depths are 4½, 5½, 6¼, 7 3/8, 9 1/8, and 11¼ inches, and boxes are offered to suit. In USA they are known as Half depth, Shallow, Illinois depth, Three quarters, Full depth, and Jumbo frames and boxes. I have seen three Illinois depth boxes in use as the brood chamber in Canada, and was told that it was not uncommon.

Guilfoyle's (Queensland, Australia) catalog of beekeeping equipment offers boxes and frames in 4 depths, viz: Full depth (9 1/8 in), W.S.P. (7 1/8 in) Ideal (5¼ in) and Half depth (4¼ in). He adds a little rhyme which is worth repeating:

At 20 young Jim was so full of vim he knew that full depths
were the size for him
At 40 with hundreds of hives of bees he thought he'd do better
with W.S.P.'s.
At 60 with old age hard on his heels his slipped disc and
strained muscles called out for Ideals
Now Junior Jim has ordered supplies to re-box the lot in the
Full Depth size.

For those willing to take the plunge, the Langstroth
Jumbo could be a good choice. It has been strongly advocated
and is in considerable use. It is a Langstroth hive of Dadant
depth (10 frames 17 5/8 x 11¼ inch) taking standard Langstroth
supers, excluders, roofs, etc. Like the standard Langstroth, the
inside width of the hive permits the use of British Standard
frames if rebates are provided in the side walls, and this could
facilitate the change for those with colonies on BS frames.

I do not consider the National or Smith hive (11 frames
14 x 8½) large enough for single box working. I know that many
beekeepers claim that it is, and say that they work strong
colonies of bees with broodnests of this size with good results. It
is not my experience. I find that the first super gets used for the
storage of pollen, and becomes part of the broodnest for much if
not all of the year. I hate pollen in supers, and I very much
dislike (because the combs are not interchangeable) a brood
chamber of one deep and one shallow box, which is what the use
of a single box of National or Smith size leads to. For me, use of
the National or Smith hive, or for that matter the Langstroth,
is, of necessity, a double brood box operation, although none the
worse for that.[6]

I have long been urged to try the so-called British Deep
frame, 14 x 12 inches, and for the last few years I have had
some colonies on this frame size. It has been about a long time,
but has never become popular. Mine are short lugged, in a extra
deep Smith box with a top bee space, and I explain later how I
use these deep box colonies. The frame is larger than a
Langstroth or a British Commercial (16x10) and is of adequate
depth to obviate pollen in supers, and of adequate size for single
box working. But at around 75 to 80 lbs. weight when filled

with brood, bees and stores, it is not a hive for single handed migratory work.

Much as some people seem to dislike it, I positively like and prefer double brood box working. I like the flexibility it provides, which greatly assists management. Some of the supposed disadvantages are imaginary, such as the view, so frequently expressed but quite unfounded, that bees winter less well on two boxes than on one. It is not so. Far from being a hindrance or a barrier, as some suggest, the gap between the boxes assists bees in their winter movements; they can move from comb to comb other than around the ends. Also that the space between the two sets of combs is repeatedly bridged with brace comb, so that the bees resent the boxes being separated, which just doesn't happen if the frames have wide top bars and the proper bee space is maintained between the two sets of combs. It seems to be forgotten that double brood box working is the normal practice for most beekeepers (using Langstroths) worldwide. It would not be so if it made management more difficult.[7]

I write from a long experience of double brood box working, mainly with brood boxes of National size, and have come to use operational practices that have, as often as not, been developed by Langstroth users. These practices reflect, and make use of, the flexibility inherent in double brood box working, largely irrespective of frame size. They are as relevant to users of National hives, as to the much smaller number of beekeepers who use Smith hives, as I do.[8]

Frame design

A most undesirable hangover from the WBC tradition is the continued use of brood frames with top bars 7/8 inch (22mm) wide. Hoffman self spacing frames are coming into increasingly common use, and replacing metal ends, which is good to see, and the use of wedge fixing for foundation gives a suitable depth of top bar, but to make the top bar 7/8 inch wide is quite indefensible, and, I suppose, done only to keep cost down. The top bar should be 1 1/16 inch (27mm) wide and nothing less. Frames with such top bars can now be had from the appliance manufacturers in BS size for a few pence extra each. I used to have to order them specially. Such top bars are standard with

Langstroth and Dadant and now also with 16x10 frames. With top bars 1 1/16 inch wide and around 11/16 inch (17mm) thick properly spaced in the brood box there will be little or no brace comb built between the top bars. With top bars 7/8 inch wide, even of the same depth, there will be a great deal. It is worth a great many pence to me to avoid such brace comb and facilitate smooth and easy operation.[9]

The bottom bars of frames should be the full width of the side bar (usually 7/8 inch) whether the bars are of two pieces or one. Full width bottom bars assist in preventing brace comb, and also guide the uncapping knife when that is necessary. I used to have to order such bottom bars specially, but they are now readily obtainable. Indeed side bars and bottom bars 1 1/16 inch wide can now be had, and could be still better.

Arthur Dines, writing in Bee Craft, said about all that needs to be said about Hoffman frames.[10] He accepted that they have some disadvantages, especially in relation to many of the (tangential) extractors in use in this country. With radial extractors there is no problem at all. He pointed out their main advantages, not least the facility for moving or lifting several combs at a time, and the almost total elimination of crushing bees during comb handling once the technique of handling is mastered; and the simple drill to keep Hoffman frames really tight together at all times. As he said, properly used there is no other spacing more positive and accurate, and none more likely to remain so during the long life of that frame. Arthur Dines added, as I have also pointed out, that the great bugbear of British hives - the lifting of the combs below when an upper box is raised - does not occur at all in hives with Hoffman frames and top bee space.[11]

Both Dr. Miller and Carl Killion used and advocated frames with top bar, side bars, and bottom bars all 1 1/8 inch (28mm) wide, nail or staple spaced ¼ inch (6mm) apart. I have seen such frames in use in Canada, and they were certainly very free from propolis (with Caucasian bees) and thus easy to part and remove. Such frames maintain the desired spacing, as Killion says. They would have to be made by the beekeeper, but would be very easy to make, and could be made from offcuts and cost very little. Brother Adam uses a frame that is similarly spaced.[12]

I like and use grooved side bars. They hold the foundation where it should be, plumb in the centre, and ensure that the bees fix it firmly to the wood. Two piece bottom bars also help in this respect. If you wire your frames, drill the holes for the wires and eyelets on one side of the groove, not in the centre, so that the foundation can be slid past the wires without difficulty before the wires are embedded.

Frame spacing

Frame spacing is a matter of some importance. The frames I use in my Smith hives have 1 11/32 inch (34mm) spacing when new, viz. a nominal 1 3/8 inch spacing, which is the standard practice with Langstroth users. I find it very satisfactory. But many Hoffman frames made in BS size are made to give 1½ inch spacing, and some users of Dadant hives also use this spacing. Despite care being taken to maintain the desired spacing, it is inevitably increased by propolisation, and 1½ inch initial spacing leads to combs being still wider apart, which I consider both unnecessary and undesirable.

I put 12 frames in a Smith brood box initially, when fitted with foundation for drawing into comb. 12 frames with 34mm spacing fit easily into the brood box and are very suitably spaced for new brood comb production. In the brood box I use 11 frames and a follower board. The use of a follower board (which I discuss below) fits in well with 1 3/8 inch spacing in Smith or National hives, and is of great assistance in maintaining the desired spacing of the frames.

Beekeepers tend to overlook, or perhaps don't know, that research work in USSR (confirmed in USA) showed that narrow spacing of combs (30-31mm or 1 3/16 ins) resulted in twenty-five per cent more brood space and a faster build up in spring. Alber measured the spacing of combs built by swarms, and found that Italian bees of Piana stock built at 30-31mm, the same as USSR found with Caucasians; Carniolans built at 33mm. Alber considered excessive spacing harmful, a cause of stress, of nosema, and of slow build up in spring.[13]

The practical problem that arises with narrow spacing is that it affords little margin for differences between comb faces. "Waviness", even if only quite slight, greatly reduces interchangeability, and drone brood on one face may mean

63

nothing at all on the opposite face. 1 3/8 inch (35mm) spacing is, in my view and that of most other beekeepers worldwide, the most satisfactory compromise.

Flat combs

Combs need to be flat to be interchangeable - and I mean flat - and frames to be truly square and hanging strictly parallel in the box. When making up frames, which should be glued and nailed, take care to ensure that they are square and parallel - that each and every one could interchange with any other without detection - and never forget that they may be in use for thirty years or more (albeit fitted with new foundation from time to time).

One can get flat combs from ready wired foundation, but unless the foundation is drawn into comb very rapidly and soon there is an unhappy tendency for such foundation to "wave" and produce slightly wavy combs. Wiring frames yourself - three or four horizontal wires electrically embedded in unwired foundation - is more certain to produce flat combs, but is an extra operation. I have wired a great many frames in my time, but nowadays I mostly use wired foundation in brood frames, and I try and ensure that it is drawn into comb rapidly and immediately after fixing by giving a box of foundation as a super at the right time.

My aim is to get perfect combs - quite flat and therefore fully interchangeable, fully drawn out and completely filling the frame, and nothing but worker cells. It can be done, and should be the aim. I cull brood combs very severely, not on account of their age, but if they do not come close enough to perfection. Work the poorer combs to the flank of the broodnest and in due course such combs will be free of brood and full of stores. Remove them, extract the honey and cut out the comb for melting down. Clean the frame and re-use it when you need.

Except over a slatted rack, foundation drawn out into comb in the brood chamber is seldom fully drawn down to the bottom bar. Foundation should be drawn out into comb in an upper box, when it will be drawn down to the bottom bar, and to the side bars too, if it is properly fixed in the frame.

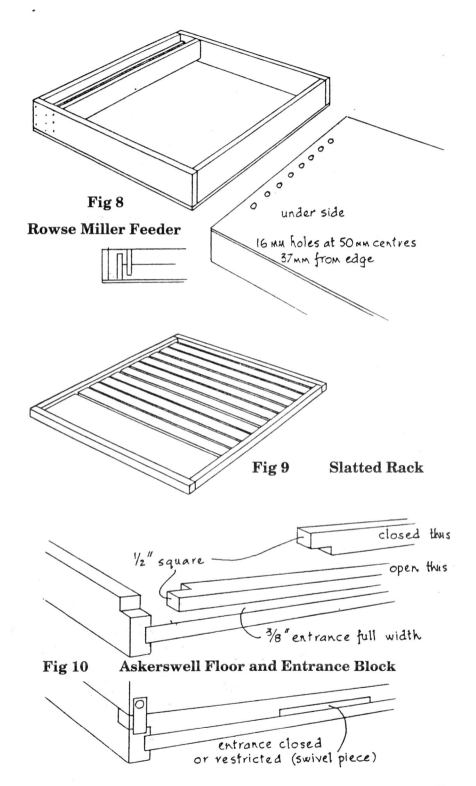

Fig 8

Rowse Miller Feeder

under side

16 mm holes at 50 mm centres
37 mm from edge

Fig 9   **Slatted Rack**

closed thus

½" square

open this

⅜" entrance full width

Fig 10   **Askerswell Floor and Entrance Block**

entrance closed
or restricted (swivel piece)

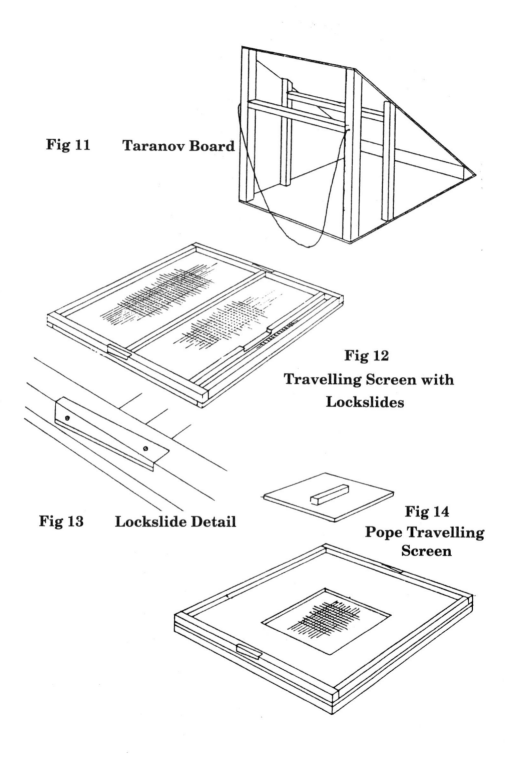

Fig 11   Taranov Board

Fig 12
Travelling Screen with
Lockslides

Fig 13   Lockslide Detail

Fig 14
Pope Travelling
Screen

Follower boards

MAFF Advisory Leaflet 445 calls follower boards spaced dummy boards. A follower board or dummy board is not a division board. It is essentially a board the same size as a brood frame, similarly suspended from the top bar, and as easily removable. In the form in which I use it both the board and the top bar are from 3/8th inch (9mm) material. On one side are two 1/8th inch strips (one with a V-edge) to match with the Hoffman frames and on the other side four small staples are put in to ensure a minimum 5/16th inch (8mm) spacing from the side wall of the hive. (See.Fig 3)[14]

Why use them? I cannot do better than summarise what Carl Killion says in "Honey in the Comb" (1951). He says: "Many beekeepers have never heard of a follower board. In our system of management the follower is a necessary part of our equipment, just as necessary as a hive cover or bottom board. The follower is used for more than one reason. The board can be removed much easier than a frame - it is never gummed to the side wall.[15] It permits the removal of the first brood frame with ease, and its later return to the hive body before the next one is removed, as the space is adequate for this purpose. One need not place a frame of brood outside the hive at any time. The board permits better ventilation and insulation to the brood. A queen does not like to lay in the two outside combs next to the wall of the hive, perhaps because of changes in temperature, with the result that we find eight frames of brood in a ten frame hive. We use nine frames and two follower boards, and get more brood in these nine combs than most beekeepers do in ten. With the follower board in use the queen will unhesitatingly lay in the outside face of the outside comb next to the follower board."

I concur, and would add that the use of a dummy board ensures that the outside face of the flank comb is not drawn out excessively, and the comb is thus the same thickness as all the other brood combs and can be used to replace any of them. Without a dummy board the outside comb is commonly bulged on the outside face and stuck fast to the side wall.
Colonies winter well with nine or ten combs and two dummy boards in each brood box, and I often winter mine in that way. I also run some colonies through summer with ten combs and two

dummy boards. They have as much brood in the ten combs as other colonies have in eleven. A frame feeder can replace two dummy boards, when required. In queen rearing, the removal of both dummy boards gives nice room for the queen cell frame.

Dummy boards can, of course, be used to similar advantage in any type of hive. Carl Killion says, in "Honey in the Comb", "In our system of management the follower board is a necessary part of our equipment, just as necessary as a hive cover or bottom board." It is also so in mine. I have one or two such boards in every brood box.

The slatted rack

Dr. C.C. Miller, and later Carl Killion, both outstandingly successful comb honey producers, used and strongly advocated a two inch deep floor board, and a removable slatted rack on the floor board (through which the bees pass to gain access to the combs above) to prevent comb being built down below the frames.[16] A similar slatted rack is easily made for use with a standard floor board.

Such a rack is merely a rim which sits under the brood box, to match up with the brood box above and the floor board below, rebated to take the slats. (See Fig 9) When used with a top bee space hive their upper surface is 5/16 inch (8mm) below the rim to provide bee space under the bottom bars, and the bottom surface is flush with the bottom of the rim. For a bottom bee space hive the rack is set the other way up. The space between slats is 5/16 inch (8mm) and the front slat is about 4 inches (100mm) wide. That's all.

A convenient width and thickness for the slats is 1 inch and 3/8 inch or thereabouts. The rim is thus about 16 mm (5/8 inch) deep. The floorboard will provide a space below the slats of ¾ or 7/8 inch. The floorboard is unaltered, and an entrance block is used in the normal way. I use the ¼ inch x 5 inch entrance that the block provides for much of the time, giving a full width entrance in summer. The rack can stay in place winter and summer or be added at dandelion blossom time and removed for winter, as you please. I leave mine in all the year round.

Slatted racks can be bought in USA and Canada, in Langstroth size, but cannot be bought anywhere in any other size, as far as I know.[17] But they are easy to make, and cost next to nothing. I have them in use under all my 14 x 12 boxes, mainly to ensure that the combs are fully used down to the bottom bar, which they are. Without a slatted rack there is a tendency for the combs not to be so fully used.

Dr. Miller, and Karl Killion, and more recently Charles Koover, were convinced that they helped to prevent swarming by allowing the colony better to control ventilation, and all three of them used them and advocated their use. I know that a number of others, including some quite large scale beekeepers, also use them. I think it likely that they do assist the colony to control ventilation, and add to the comfort of the colony in adverse conditions. It is not really possible to say.

Entrance block

An ingenious but simple and effective entrance block is shown in Fig 10. It is available from at least one manufacturer as the "Askerswell" block. The arrangement easily and simply permits a full width entrance 7/8 inch or 3/8 inch deep; a restricted entrance up to 4 inches wide; or complete closure, e.g. for travelling; and an "easy fit" block secure when in place but not requiring a hammer to bang it into place for travelling or a screwdriver to prise it out again. The design of block is suitable for any single walled hive.[18]

Lockslides

I discovered Lockslides in use on Ernie Pope's Yarrow hives in 1963 and found that many other Tyneside beekeepers were also using them. They consist of an anodised metal rail, fixed with the aid of a template in a precise position near the top and near the bottom of two opposite sides of every box, so that the boxes can be held together by a removable wedge shaped slide.[19] (See Fig 28). A tap with the hive tool or pressure with the thumb and the slide is removed and the box can be lifted away. Push the slide on firmly with the hive tool and all is securely fixed.

To ensure that the rails are fixed in the correct place, so that the wedge shaped slides always fit, a template appropriate to the hive on which they are to be fixed is supplied with the lockslides, when bought; and the slides are available in two widths so that framed excluders can be accommodated by using the wider slide.

For beekeepers with a small number of hives they are ideal. For me it meant screwing or rivetting about eight hundred rails in position, and I hesitated a long time, and tried a few, before taking the plunge. But the advantages and general usefulness of the device were so evident that I eventually did so, and by about 1968 lockslides had been fixed to all my floors, deep and shallow boxes, and travelling screens.

With lockslides on all my boxes, etc., hive staples and strapping are no longer required for transporting hives, and floorboards and screens are securely fixed but can be released in an instant. The spring floorboard change is much easier, and made with little disturbance to the colony, and when parting two brood boxes to inspect for queen cells, the upper brood box and super(s) remain as one piece and are parted only if circumstances require.

Travelling screens.

Travelling screens are not difficult to make but they should be accurately made to the correct dimensions, and made of good material that is unlikely to warp or twist. I use 8-mesh hardware cloth, not perforated zinc, fixed between a 6 or 7 mm thick rim below and a 22 mm thick rim above, with a 22 mm thick piece across the middle from side the side, which both adds rigidity and permits and facilitates strapping. (See illustration, Fig 12). The design will allow the entrance block to be stored on the screen after removal, and will allow the colony to be left screened, with cover board and roof on top, at least until the next visit after moving.

All my screens are fitted with lockslides, as shown. They greatly facilitate closure, release, and re-fastening. The wire mesh cloth should be of fairly heavy gauge, and thus not easily damaged or prone to rust. It can be brushed over with black bituminous paint after some years' service, to further prolong its life, if need be.

Ernie Pope's screens (which also have lockslides) have a central square of wire mesh of about half the area of an otherwise solid board of 9mm ply. He covers the screen with the

70

piece cut out of the board when the colonies are on the new site. (See Fig 14) It has to fit the square, so both the loose piece and the board from which it was cut carry identification by letter or number. Preferably, the hole should be accurately cut to a standard size for all screens, so that any of the pieces cut out will fit any screen.

This type of screen forms an adequate inner cover when the cut out piece is replaced, and is very suitable for moving colonies to and from the heather, when nights (and early mornings) are generally cool. For moving strong colonies at other times of year I prefer to have the maximum area of screen mesh over the colony.

The Taranov Board

With the aid of a Taranov Board an artificial swarm can be made from a colony found to have advanced queen cells, i.e. at a stage just before it would naturally swarm if left alone, that will behave in all respects as a natural swarm. The bees will stay on the new site when hived, even if that site is close to the old site, and will work with the same prodigious energy as a natural swarm. I discuss the use of the board in Chapter 8.

A Taranov Board (See Fig 11) is very simple to make. Two pieces of exterior grade hardboard form (1) a base board about 18 x 15 inches (450mm x 380mm) and (2) a sloping board about 20 x 15 inches (500mm x 380mm) - the size is not critical - glued and nailed to (3) a batten about 1¼ x 1¼ x 15 inches (32mm x 32mm x 380mm) ripped along its length to about 40 degrees, and to (4) two supporting pieces at the open end, so that one of the narrower edges of the sloping board is approximately at the height of the hive entrance while the other is on the ground. Fix a cross bar, or two, to the two supports, both to make the construction rigid and to provide something around which the bees can cluster. Attach a length of cord to facilitate carrying.

Smokers

A really effective smoker is an essential tool for the beekeeper. What is required is a smoker that will produce plenty of smoke when it is needed, will burn for a long time without having to be refilled, (and all day if refilled from time to time), is not too heavy, yet substantial enough to last for years. The bent-nosed smoker of the Root or Dadant pattern is the only type worth considering. (See Fig 26) It can be had in tinplate, in stainless steel or in copper, and should have a good sized barrel, say about 7 inches x 4 inches, or more.

I have three smokers, two of stainless steel and one of copper. One, 10 x 4 inches, does a fine job, but I don't get on with it too well. I find it a bit heavy and cumbersome. The other two, one stainless steel and one copper, both 7 x 4, and both with heat shields, are more or less interchangeable. I think you need more than one, just in case. You can't do much without a smoker! I have had the copper smoker (a Woodman, and a great favourite with me) for more than fifty years. It has had all the copper work renewed (once) and the bellows have been replaced several times. In fact all that remains of the original smoker are the two bellows plates and the heat shield.

I like to have a heat shield on my smokers. I long ago got into the habit (perhaps a bad one) of gripping the smoker between my knees. I also have a hook attached to the back plate, so that I can hang it over the hive wall, and two small straps fixed to the front plate in which I keep a Root type hive tool. (See Fig 26).

Smokers that are used much have to be cleaned out from time to time, or they get clogged with carbon deposit, both at the air intake and at the nozzle. It pays to keep smokers in really good working order.

Half-rotten sacking is a good fuel, but not so readily obtainable as it used to be. Planer shavings are also good, and easy to get. Rotten wood, e.g. from a fallen tree, can be carried home in large pieces and broken into small pieces for use in the smoker, where it is an excellent fuel. Dried pine cones are good, and dried (mown) grass can be used. Take sufficient fuel with you when you visit colonies away from home, and take a small bag of wood shavings or a supply of tissue paper to start the smoker going.

72

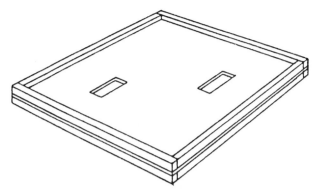

Fig 15   Clearer Board for Porter Escapes

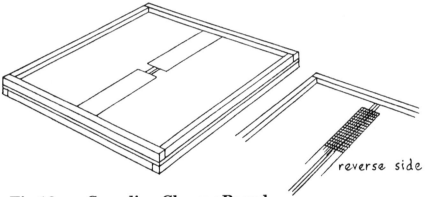

reverse side

Fig 16   Canadian Clearer Board

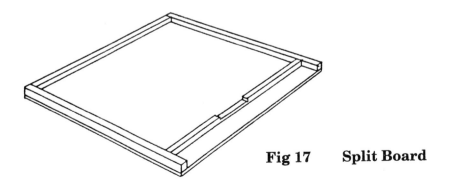

Fig 17   Split Board

Queen Introduction Cages

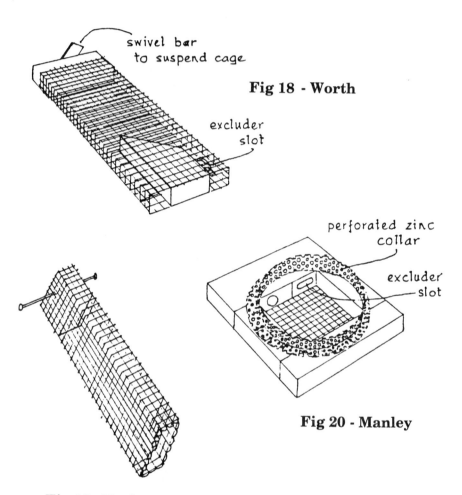

swivel bar
to suspend cage

Fig 18 - Worth

excluder
slot

perforated zinc
collar

excluder
slot

Fig 20 - Manley

Fig 19 - Butler

Foundation Strips for Miller Method

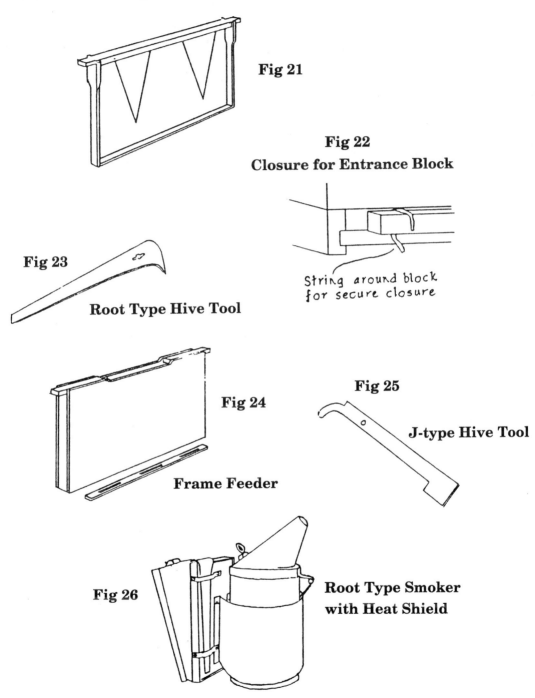

Fig 21

Fig 22
Closure for Entrance Block

Fig 23

Root Type Hive Tool

String around block
for secure closure

Fig 24

Fig 25

J-type Hive Tool

Frame Feeder

Fig 26

**Root Type Smoker
with Heat Shield**

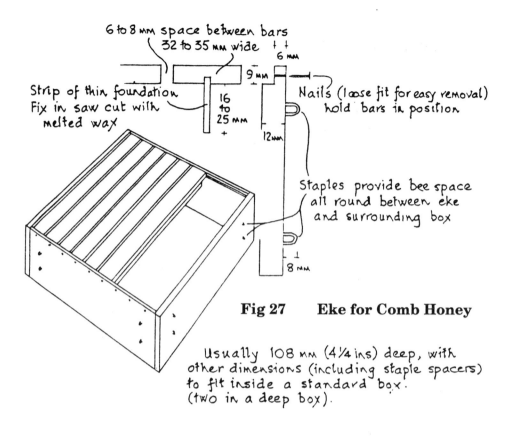

6 to 8 mm space between bars
32 to 35 mm wide

6 mm

9 mm

Strip of thin foundation
Fix in saw cut with
melted wax

16
to
25 mm
+

Nails (loose fit for easy removal)
hold bars in position

12 mm

Staples provide bee space
all round between eke
and surrounding box

8 mm

Fig 27   Eke for Comb Honey

Usually 108 mm (4¼ ins) deep, with
other dimensions (including staple spacers)
to fit inside a standard box.
(two in a deep box).

Fig 28   Lockslides

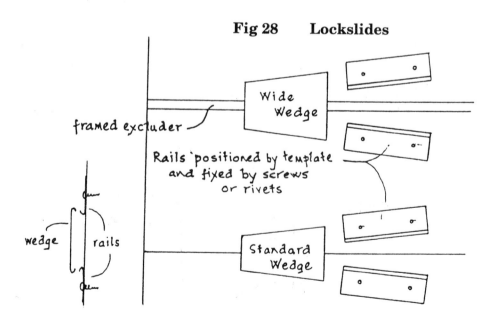

Wide
Wedge

framed excluder

Rails positioned by template
and fixed by screws
or rivets

wedge   rails

Standard
Wedge

Queen excluders

Queen excluders are an essential part of my outfit and operation, as I expect they are for others. Excluders made from sheet material, zinc or plastic, need to be framed (at least for top bee space hives) and to be supported by cross strips. I have a good many, but I prefer wire excluders. I don't find Waldron excluders the nuisance with brace comb that some other beekeepers seem to, perhaps because of both the top bee space and the wide frame top bars, but the wire excluders that are set flush with one side of the surrounding frame, and provide a bee space on the other, as the B & J and the Herzog do, are certainly preferable.

Excluders should be cleaned after use, and kept as clean as possible. A solar extractor, large enough to take excluders, does quite a good job with wax, but I first use the hive tool as a scraper for the removal of propolis on the frames and finally a wire brush to clean them up.

The precise width of the slots, or of the space between the wires, is what excludes the queen, and care needs to be taken to avoid damage or distortion such as to allow a queen to pass through. For that reason I seldom use a blowtorch to clean excluders.

Supers

In BS size, 14 x 5½ inch frame, I find I need to have about five supers for every two colonies.

I have several supers to take 14 x 4¼ inch frames, which I converted from section racks twenty years ago and cut down frame side bars to suit. Fitted with extra thin foundation they are ideal for cut comb and are used for that purpose. Most radial extractors won't take such shallow frames, if honey has to be extracted from them, but my MG Parallel Radial extractor will.

All the frames in the supers, without exception, are Manley frames, spaced at 1 5/8 inch, with wedge top bars and grooved side bars. They take wired foundation for extracted honey, or extra thin (and, of course, unwired) foundation for cut comb, with equal effectiveness. The top bars and bottom bars are of the same width (1 1/8 inch) and are used as a knife guide in uncapping. You need a radial extractor to extract honey from Manley frames. Given that, they are exactly what is needed.

As Manley found, 1 5/8 inches is just about the maximum spacing at which foundation will be drawn without a good deal of wild comb being built between the sheets of foundation. Fortunately, ten Manley frames with 1 5/8 inch side bars fit very conveniently in a Smith or National super, which have the same internal measurements, and in Dadant supers (which Manley used). But don't allow propolis to build up much between the side bars or it will be difficult to get the frames out of the super.

Nine frames with this width of side bars (1 5/8 inch) just won't fit in a Langstroth super, and Manley frames for Langstroth supers are slightly less wide (nine to a super).

My supers are used for honey and for nothing else. Good extracting combs, built from thick, wired, foundation, well secured in the middle of the frame, will be good for many years, and are a valuable asset. Because they have never been bred in, wax moth is no problem. I try hard to avoid pollen in super combs. Only by accident, and that very rarely, do any get bred in.

Frames should be made up with great care and complete accuracy, so that any one could replace any other without detection. Properly assembled, nailed and glued, such frames will be satisfactory for forty years or more.

Fume boards

It is sometimes convenient to clear bees from supers with a repellent, such as benzaldehyde (artificial oil of almonds) or propionic anhydride, applied on a fume board.

A fume board is very simple to make. It is merely a wooden rim, one or two inches deep, to match the dimensions of the super, to which a piece of cloth, or felt, or other soft material is fixed. I use a piece of soft particle board (Celotex) and cover it (on what will be the top) with printers' aluminium sheet, painted black. The repellent is sprinkled on the board with the rim upwards and the board is then inverted and placed on top of the super.

Wet the fume board with about a tablespoonful of benzaldehyde sprinkled around. Smoke the bees down half an inch or so, then place the fume board squarely on the super. Look under the fume board about 30 seconds after it is first put in place. If the bees are not moving and appear dazed they are being subjected to too much repellent and the fume board should be placed crossways on the super.

Only experience will teach effective use of fume boards. They work better if covered with metal and painted black to absorb the sun's rays and get warm. On a warm day, with the temperature around 70 degrees F, most of the bees will be driven out of a super in ten minutes or less. In cooler weather it takes longer. A shallow rim, and not too much repellent, will drive the bees down about six inches, which is what I want. Few boards are needed (I have four) as they are moved from hive to hive as required.

Hive tools

A really good hive tool is a necessity, and as Manley says "quite essential for use with the single walled hives used on bee farms". In the forty years or so since he wrote this (in "Honey Farming") good hive tools have become readily available from bee appliance suppliers, but some are much to be preferred to others. They must, as Manley says, be made of high quality steel so as to be light and at the same time exceedingly strong. Those offered in stainless steel are very good, but I like mine rather shorter (about 7 inches long rather than 9 inches) and have them made to the length I want.

If you are working on your own it is often helpful to use two. I have several, of two distinct types. One, a 7 inch stainless steel one, of the Root type, is housed in a pocket on the smoker, and lives there. I have a similar one in the bee box and another at home. They are excellent for cleaning equipment as well as for use working with the bees.

In about 1949 my old friend H.C. Crowther (an ingenious engineer/beekeeper who invented an improved Porter type bee escape now commonly available) gave me quite a different type of hive tool that he had brought back from New Zealand - the so-called J-type. I don't know whether it originated there. Crowther had a number of copies made for friends, and I took several with me to Devon in 1955, and gave some away. Later, in Northumberland, Ernie Pope had several made in stainless steel, and we gave some of those away. Nowadays the J-type hive tool can be bought from any of the suppliers of bee goods, and are deservedly popular.

A hive tool is required first to exert leverage to prise boxes and frames apart, and second to assist in prising the frame out of the box. This last use is where the J-type hive tool is so useful. The hive tool is, of course, also used for scraping and cleaning up and killing queen cells or drone brood.

The Root type hive tool has both ends sharpened and has the broad end, about 1½ inches wide, turned over at a right angle for use as a scraper, and the narrow end, about ¾ inch wide, for use as a prise or lever. The J-type has a blade with a sharpened edge for scraping and exerting leverage at one end, but at the other a shoulder and blunt hook with which one can exert leverage upwards under the short lugs and lift the frame for removal from the hive. I use both types, but for routine inspections of bee colonies it is the J-type I immediately pick up. (See Fig 23 & Fig 25)

Wedmore points out that the taper at the sharpened end must be right, and suggests a taper of 1 in 8 and an end nearly sharp.[20] The taper needs to be right to facilitate parting the boxes without damage to the wood, and a nearly sharp edge facilitates scraping.

Uncapping fork

An uncapping fork (Fig 37) is a useful tool. I use it as an adjunct to an uncapping knife, not instead of the knife, as I believe some people do. With Manley frames and well filled combs a sharp Granton knife, used cold, is the best uncapping tool I know. But the uncapping fork will lift the cappings from less well filled combs and from area that the knife does not reach, and do so most effectively.

I have also found the uncapping fork to be a very effective tool for scraping partially granulated combs to the midrib. For honey from oil seed rape, granulated solid and set hard, an old tablespoon with a sharp edge is much the best scraping tool, but if the granulation is less complete and set less hard the uncapping fork is the better tool. Scraping heather

80

honey to the midrib for pressing can also be done quite effectively with an uncapping fork, but for this job the Smith cutter/scraper (Fig 36) supplemented by a tablespoon, is best.

I carry an uncapping fork in my bee box these days; it is quite the best tool for uncapping combs in the hive, whether of brood or of honey, and much better than a hive tool, which I formerly used. It is good practice to encourage bees to make use in spring and summer of patches of overwintered stores, which they will do if the cappings are broken. Such patches of stores are commonly granulated in the comb. The uncapping fork, with its closely spaced prongs, is the ideal tool for this purpose. It is also quite the best tool to use to uncap drone brood in varroa control, or at other times.

Scissors

In my bee box I keep a very good small pair of scissors, with small blades and blunt tips, for clipping queens, and for that purpose only. They are used for nothing else. The type I use are "baby scissors" designed and sold for cutting the nails of babies, I believe.

Veils and overalls

A veil is a necessity, and should always be worn. As Manley said, (in Bee Craft) "No sensible person will open hives unveiled, unless one is really experienced, and can assess conditions, etc." I always wear a veil; there is no sense in risking being stung about the face and eyes.

I have long since got used to a very simple combined hat and veil, and find anything else less comfortable. It blows about a bit, but it is very light, and folds flat, so that I always have one available, in the car, even when I'm not on bee business. But there are some excellent veils to be had, these days, and very good combined bee suits and veils. I have, and frequently use, one of these.

Overalls, or a bee suit, are, I suppose, not essential, but a good quality white bee suit in cotton or a cotton mixture (not in nylon) is a very convenient garment, and I always wear one. It has pockets into which I stuff pieces of torn up sacking for smoker fuel, perhaps gloves, and one in which I can put a hive tool, etc. It gets surprisingly dirty, so must help to keep ones clothes clean. I stuff the trouser bottoms into short rubber boots, to avoid stings around the ankles, or bees crawling up.

Gloves

Without hesitation I advise the purchase of a good pair of bee gloves. Soft leather gloves with cloth elbow-length gauntlets are what is required. A really good pair is quite expensive, but much more convenient in use than cheaper substitutes, and a pair will usually last a small scale beekeeper several seasons with proper care.

In a good purpose-made bee suit and veil the occupant wearing gloves is almost impervious to bee stings, and is meant to be so. But I do wonder whether the beekeeper who invariably wears such a suit, plus gloves, becomes rather too insensitive to bad temper in colonies. Do handle bees without gloves whenever you can, and wear gloves only when you must. I keep a pair of gloves in the pocket of my bee suit and wear them when I need, such as to ensure clean hands when I expect to handle a queen. Wearing gloves routinely means insensitivity to temper in bees, which is most undesirable.

Clearer boards

I have both Porter escape boards and Canadian type clearer boards in considerable numbers, and use both types, but I don't consider one better than the other, or even as being equally suitable for the job. Each type is to be preferred, in my experience, in different circumstances, which is why I have both. I will try and explain.

The Canadian type (Fig 16) has no moving parts and does a remarkably quick job of clearing bees down. That is what it was designed to do - to clear supers quickly. I find that when bees are flying well, most of the bees will have left the supers in an hour or two after inserting the clearer board, but I also find that the supers will almost always retain some bees no matter how long they are left above the board. Possibly some bees find their way back - there is nothing to stop them - but of course one doesn't know.

In contrast, a clearer board fitted with Porter escapes works far more slowly, but completely effectively (provided the escapes are in good order). The usual practice is to put Porter escape boards on in the evening and remove the supers 24 or 48 or 72 hours later, by which time - certainly by the later time - the supers are completely clear of bees. They can't get back, once they have gone down.

But the use of Porter escapes requires two visits, one to put the boards under the supers, and the second, 24 or 48 hours later, to take the supers off. For someone like myself, with all but two or three of my colonies in out-apiaries, often a considerable distance from home, the Porter escape, for all its effectiveness in removing bees from supers, given time, is not wholly satisfactory. It is also quite impracticable for use in removing supers of honey from oil seed rape, as you will almost certainly find that the honey has begun to granulate in the supers when you collect them.

At the rape, bees have either to be shaken off the super combs, or cleared from the supers by fume boards and benzaldehyde, or by Canadian type clearer boards, so that the supers can be brought home and extracted that same day (continuing into the night if need be). I use all three methods as I think appropriate at the time.[21]

The value of the Canadian type clearer board is that it will clear eighty per cent of the bees from the supers in two day-time hours, making it easy to get the relatively few remaining bees out of the supers by shaking or by stacking them on the trailer and allowing them to fly back.

Vast areas of oil seed rape are grown in Alberta, Canada, and I was told there (and shown) that the so-called Canadian clearer boards (of which there are several types) were designed with precisely this purpose in mind.[22] That is how and when I use them.

The pattern that I use consists of a 9mm ply board, (it should not be less thick) with a slot about half an inch (12mm) wide across the full width. This slot is covered on the upper side, except for an inch or two in the middle, by a metal strip, and is covered on the lower side with wire mesh, except for an inch or so at each side. Mine have a 5/8 inch (16 mm) thick rim below and above the board. (See illustration, Fig 16). The board appears to work equally well either way up. It has the merit of permitting drones to pass down, which Porter escapes may not, and although one should not have drones in supers it does sometimes happen, and can cause the escapes to get blocked.

At the end of the season I use clearer boards with Porter escapes, as I can then find time (and a use) for a second visit and one doesn't have to be in a rush to extract. But my design of clearer board for use with Porter escapes differs in two respects from those commonly available. First, notwithstanding that my hives have top bee space, my Porter escape clearer boards have a 5/16 inch (8 mm) thick rim below the board and a 7/8 inch (22 mm) thick rim around the upper side. These rims ensure that there is space above and below the combs and no chance of the escapes being blocked by comb built above or below the frame bars. And second, since I may use these boards as feed boards after removing the escapes, I ensure that the two holes into which the escapes fit are so positioned that they can each give access to one of two 1 gallon contact feeders if I choose to give two (see Fig 15). Such clearer boards, given proper attention to the escapes, usually work perfectly.

A Porter escape board can also double as a split board, if an entrance that can be opened or closed at will is cut in the rim to provide an entrance to a colony above when that is needed. Of course, the entrance must be securely closed when the board is in use as a clearer board.

Porter Escapes

The essence of the thing is that they should work. Sometimes they don't, and the fault usually lies with the beekeeper and with inadequate attention to detail. The springs should nearly meet, but not quite. A gap of 1/8 inch is about right. The tunnel should be clean, free of propolis and free of obstructions. It should be possible (but it often is not) to separate the two pieces of the Porter escape. The springs are best adjusted while the cover is removed. A smear of vaseline before re-assembly is worthwhile.

Never leave Porter escapes in place longer than needed; the bees gum them up with propolis very quickly. They can be cleaned in one of two ways. If they are not too bad, put them in a screw cap glass jar, cover with denatured alcohol (Boots), cap

84

the jar and leave for twenty-four hours. Then remove them and use an old toothbrush and warm soapy water to brush them clean. Save the jar and the alcohol for another batch; it can be poured through coffee filter to remove dirt and debris if need be. If they are very gummed up the best way to clean them is to put them in a strong solution (about a tablespoonfull to a pint) of washing soda in boiling water. Whether they are metal or plastic, boil them in the solution for a while.

Cover boards

It is customary, it would seem, and certainly commonly advised, for inner cover boards to double as clearer boards, and feed holes in cover boards are usually made to take Porter escapes. But cover boards are likely to be in use as cover boards when escape boards are needed. I leave the inner cover boards in position on top of the hives and use clearer boards designed for the purpose, which sometimes double as split boards with an entrance in the rim to a colony above at other times of the year.

Cover boards (which, for my hives, have no rim on the under side, and a 5/16 inch thick rim above) should have two holes for feeding, positioned so as to permit access to two 1 gallon contact feeders. (See Fig 6).

Feeders

I discuss feeders and feeding in another Chapter. Here I need say only that I have, and use, three types of feeder, as follows:

1.   Manley or Miller type box feeders, mainly of the design used by David Rowse (the Rowse Miller feeder) to hold two gallons of syrup, (see illustration, Fig 8), one for each colony. They take up a lot of room in store and most of them double as cover boards on the hives through the winter. The materials and construction should take this into account. With colonies in out apiaries one needs feeders that permit two gallons of feed to be given to each colony at one visit, and for this purpose the Rowse Miller feeder is excellent.

2.   About half that number of one gallon plastic bucket feeders, with a fine copper gauze panel in each lid through which the syrup is taken. Such feeders can be bought from the appliance manufacturers. They are very useful, and take up little room in store as they stack inside each other with the lids removed.

3.   A few frame feeders for feeding nuclei. (Fig 24).

I also have a number of 3 gallon home wine makers' fermenting bins in which I make syrup and carry it to out apiaries. And perhaps I should add that I have about 50 one gallon lever lid paint cans with perforated lids that have seen many years service as feeders, and are still serviceable and occasionally put into use.

Queen introduction cages

The Butler cage is well known and widely used. As I use it, the cage is suspended by frame nails (see Fig 19) between two combs, for which the removal of the follower board leaves ample room. I sometimes use the Worth cage (see Fig 18), which is similarly suspended, and has a facility for access to the queen by worker bees through an excluder slot and the delayed release of the queen until a small piece of candy in a second slot is consumed. I cover the open end with a small piece of paper, secured by a rubber band, as I do with the Butler cage, and omit the candy. The Fileul cage was very similar to the Worth cage, and could also be used as a postal cage, but seems no longer to be available.

This principle of delayed release was inherent in the Manley cage (see Fig 20) in which the queen is caged on the comb. I still have in use the six Manley cages that I made in 1947. But Manley cages cannot be bought, as far as I know, and are not easy to make.[23]

All three kinds of cages (and the Fileul) are intended for the introduction of queens unaccompanied by worker bees. I tend to use Manley cages in preference to Butler or Worth cages, but I can't say that I have found any difference in their effectiveness. All three are very satisfactory.

5   Colony Records

I can tell more of less of the history of every colony
of bees since I began keeping bees in 1861.

Dr.C.C.Miller, "Fifty Years Among the Bees" (1911)

My record book

In my business life a time planner diary has long been a
necessity, as it is for most professional people. The form of desk
and pocket diary that I use is described by the manufacturers as
"a diary and activity book and a work and project planner." This
is an apt description of my hive record book.

I suppose most beekeepers keep some sort of colony
record. It would be foolish not to do so. I have seen brief notes
on pieces of card kept loose on the cover board or pinned to it or
to the underside of the roof. Some beekeepers use a broken
piece of section to make notes on, sometimes pinned to the back
of the hive so that it can be read without disturbance to the
colony. Some use a loose-leaf book of hive cards. Both the
BBKA and the BIBBA have produced colony record cards for use
by members, and other forms of record cards can be bought from
appliance manufacturers. None of them, and none of these
practices, suit my purpose.[1] I make my notes in a record book,
which lives in my bee box and is thus always with me when I am
with the bees.

No doubt other beekeepers have devised and use
satisfactory systems of keeping colony records. Some may be
better than mine. They are seldom described by authors of

books on beekeeping, or by writers of articles in the bee press. An exception is to be found in Kenneth Clark's excellent Penguin handbook "Beekeeping", in which he describes his form of notes and his "shorthand" symbols.[2] He used pieces of sections, kept under the roof or clipped to the back of the hive, on which to make his notes.

A beekeeper with two or three colonies can make notes as full as he pleases. Some appear to rely on memory and make no note at all. For the man with ten or more colonies some form of shorthand is essential, or the drudgery of writing will lead to shortened and inadequate notes, or to no note at all. In my view it should be standard practice to make a full note immediately the beekeeper finishes with hive A and before he starts on hive B. It soon becomes a habit.

In the late 1940's I developed a form of shorthand for making my notes, and have used this unchanged, except for adding a few symbols, for almost fifty years. It is not original, as it was closely modelled on the system devised and used by Dr. C. C. Miller, and described in his delightful and instructive book "Fifty Years among the bees", (a reprint of which has recently become available). I adopted the system in 1946 and have records of every colony and every year since then. Dr. Miller, an outstandingly successful American comb honey producer (and an entertaining and instructive writer on beekeeping) in the early 1900's, considered that the winter study of his colony records contributed significantly to his success.[3] I don't doubt it. It certainly has to mine.

I explained my use of the record book and my system of shorthand notes in an article in The British Bee Journal in September 1955, and used an extract from my 1953 record book to illustrate its use. Perhaps I should add that the extract was selected for its suitability as an illustration. It is unusually comprehensive; most colony records are much briefer. This article was reprinted in The British Bee Journal in the winter of 1972, and in The Scottish Beekeeper in 1977.[4] I cannot say whether many (or indeed any) other beekeepers have adopted the system, but I can say that it is quick and easy, and very satisfactory, and will be found so by others who use it. My own beekeeping would have been much less successful without it.

My colony records are kept in a book, about 8 inches by 5 inches, of 96 pages, 25 lines to a page. Such a book can be bought at any good stationers, but, since it will get a lot of use over a period of years, it is worth paying a little more for one that is really well bound. Between the two middle pages and around the back is tied a piece of string of sufficient length to allow about 18 inches over, and at the end of this a pencil with the string tied round a small notch near the unsharpened end. Whenever you take hold of the record book you then have the means of writing in it already to hand. The book is permanently kept in my bee box. There was a time when my book lasted four seasons, and then for three, but a 40 colony unit requires a new book every two years.

Each of my colonies is numbered, a metal number tag being firmly fixed to the hive, and each colony is allotted one opening of the record book, i.e. two facing pages. Only very occasionally will more than one page be necessary for one colony in one season, but it would be unwise not to allow for the possibility, and it is not desirable to have more than one colony record in view at a single opening.

Each page has a margin ruled at the left hand edge about one inch wide, and in the margin on the left hand page I record, initially, the year of birth of the queen, her ancestry (i.e. daughter of which breeder), whether she is clipped, and relevant details of her colony's performance in past seasons, if any. In this margin I also record, if appropriate, such comments as "pure mated", "bees cross", "propolise", etc., and, as I shall explain later, records of surplus taken or feed given.

Shorthand

I shall follow an example through the season, but first it is necessary to explain the "shorthand" used. It is very simple and really very obvious, but it is necessary for each symbol to have an unmistakable meaning, and it is therefore desirable to write it down in the form of a "dictionary" until the symbols and their meaning are perfectly well known. Here is the "dictionary".

s = saw or seen	g = give
q = queen	t = take

vq = virgin queen
qc = queen cell or cells
em = emerged
eg = eggs
cap = capped
br = brood
br all st = br all stages
y = young
bs = bees
adh = adhering
sw = swarm or swarmed
art = artificial
cl = clipped
c = comb
dr = drawn or drawing
f = foundation
X = excluder
fb = follower board
divb = division board

k = kill
p = put
nst = new stand
ost = old stand
scr = screen or screened
incr = increase
D = deep box
S = shallow box
sec = section rack
rsec = round sections
emp = empty
osr = oil seed rape
occ = occupied
da = daughter
B = breeder
F = feed
FuB = Fumidil B
m/sal = methyl salicylate
H = heather e.g. S30H

= = split board i.e. with upper entrance
S½ = shallow box half full of honey
D40 = deep box containing approx. 40 lbs. of honey
cl q = clipped the queen, i.e. the action of clipping
q cl = the queen is clipped, i.e. on a previous visit
3 br = 3 combs each at least half full of brood
br in 3 = brood in 3 combs, some less than half full
DDXS = two brood boxes with excluder and shallow above
DD to DDXS = gave excluder and a shallow above two deep
gXSdc = gave an excluder and a shallow of drawn comb
4(6) = current colony No.4 (was No.6 last year)

Now we can follow a colony record through the season. First I set out the actual notes recorded at the time, (on the left hand page of the record book unless and until the left hand page is full), and then follow with the "translation" of these notes. Note that while the example is an actual one it has been selected to illustrate the use of most of the symbols and not to show a typical colony record, which is usually somewhat shorter. Note also that the colony record (Colony No.4 in the example) follows the queen (No.4) and not necessarily the hive.

90

4(6)

50	Apl. 6 Move to orch
d50B	
q cl	Apl.12 gXSdc now DDXS
51/noc/47	
52/art sw	Apl.25 s cl q br in 9 up eg in 1 d
80+incr	reverse boxes g m/sal

May 10 6 br up S½

May 19 sq kqc all st tDDwq to nst leave 1 br
w cap qc + 2c ybs + 8 empty c in new D

+28   on ost + XS28 + bs.   Incr No.15.

May 28 s q s eg & y br no qc gXSdc&f DDXS

June13 no qc put q d & X d screen DXDS

June14 move to clover

June21 no qc up or d S occ f dr

July 1 sq k eg D40 S5

July11 no qc s

July19 sq s eg & y br q lay well 1 qc cap & 1
nearly ripe no other qc so leave it.

53	Aug. 9 s uncl q lay s eg & y br br in 9 S25
gd50B	
	Oct.10 t s t X 40 up 15 d br in 2 now DD
	move to orch
+21	Oct.12 S t 10/10 g 21 lb
T49F12	Oct.17 g box F 12 lb.

Translation

The page heading shows that the record relates to colony No.4 of 1953, which was colony No.6 in 1952. The side notes show that the colony is headed by a clipped queen born in 1950, a daughter of the 1950 breeder. In 1951 her colony put up no queen cells, and gave 47 lbs. surplus. In 1952 her colony was artificially swarmed, and gave 80 lbs. surplus, plus the increase. The 1953 record shows:

Ap. 6 Move to orch   Moved from winter site to orchards

Ap.12 gXSdc now DDXS   Gave an excluder+ a shallow super of drawn comb

Ap.25 sclq br in 9up
    eg in 1d rev box
    g m/sal l

Saw clipped queen, i.e. the queen previously clipped. Brood in 9 combs in the upper box, eggs in 1 comb in the lower box. Reverse the two brood boxes. Gave a bottle of methylsalicitate.

Ma 10 6 br up S½

6 combs of brood in the upper box. The shallow super is half full.

Ma 19 sq kqc all st
    tDDwq to nst
    leave 1br w cap qc
    + 2c ybs + 8 emp c
    in new D on ost
    + XS28 + bs.
    Incr. No.15

Saw the queen. Killed queen cells in all stages. Took to a new stand the brood chamber with the queen. Put a new deep box on old stand and gave a comb with a capped q. cell and young bees from two other combs and empty combs to fill the box. Added the excluder and super containing about 28 lbs. honey with adhering bees. (Credit for the honey is given in the margin) The increase (the new box with queen cell) is colony No.15. The record follows queen No.4.

Ma 28 sq seg & ybr
  no qc gXSdc&f

Saw the queen, + eggs and young brood. No queen cells. Gave an excluder and shallow super of drawn comb and foundation.

Jn.13 no qc pqd & Xd
  gscr now DXDS

No queen cells. Put the queen in the lower brood box and put the excluder between this and the upper broodbox.
Gave a travelling screen.

Jn.14 Move to clover

Move to clover.

Jn.21 No qc up or d
  S occ f dr

No queen cells built above the excluder, or below. Super occupied but not much in it.
The foundation has been drawn.

Jy 1 sq k eg D40 S5

Saw the queen. Killed eggs in queen cells. The former upper brood box now contains about 40 lbs. honey and the shallow super about 5 lbs.

Jy.11 no qc s

No queen cells seen.

Jy.19 sq s eg & ybr
  q lay well 1 qc
  cap & near ripe
  no other qc
  so leave it

Saw queen Saw eggs and young brood. Queen is laying well One and one only queen cell and that capped and nearly ripe, so it is left.

Au. 9 s uncl q lay
  seg & ybr br in 9
  S25

Saw an unclipped queen laying. Saw eggs and young brood. Brood in 9 combs. Super contains about 25 lbs. honey. (The new queen is noted in the margin as a 1953 queen, a grand-daughter of the 1950 breeder)

Oc.10 tS tX 40up 15d br in 2 now DD move to orch	Took the super and the excluder. About 40 lbs. honey in the upper brood box and 15 lbs. in the lower box. Brood in 2 combs. Now reduced to a brood chamber of two deep boxes. Moved back to the orchards for the winter.
Oc.12 S t 10/10 g 21	The shallow super taken on Oct.10 has been extracted and gave 21 lbs. of honey. Credit is given in the margin.
Oc.17 g box F 12	Gave a box feeder of the Manley type and a feed of syrup containing 12 lbs of sugar.
T49F12	Total crop 49 lbs Sugar fed 12 lbs

Use of the records

My record book is an essential part of my day to day management during the busy season. A glance at the colony history before opening the colony is for me an essential first step, a note before moving on to the next colony is a long established habit, and a careful study of the book before making the visit ensures that I take with me to the out-apiaries what I am likely to need there.

I make a note about weather, times of flowering of important nectar sources, etc., on a page opening preceding the colony entries, and at the end of the year I record the yield of each colony and the average yield per colony (winter count) in the form shown below, on a page opening after the last colony record. The yield from each colony at the heather, and the colony average at the heather, are shown separately, so that this can readily be seen. Heather honey has to be handled separately, and differently, from other honeys, and not every colony is taken to the moor,

1953 Honey Crop

No.	Flower	Heather	Total
1	14	-	14
18 (incr from 1)	-	42	42
2	27	-	27
3	29	-	29
4	49	-	49
15 (incr from 4)	15	51	66
etc., etc.			
	538	148	688

19 autumn count)538(=28 lbs av. + Heather 6)148(=25 lbs av.

The record books show, for example, that 1955 was a marvellous year, the like of which I don't expect to see again; that 1959 and 1969 were very good and 1961 also a good year, and that in another year 12 good well provided colonies almost died of starvation on Dartmoor when the heather failed.[5]

I am credited with a good memory, but the record books are a salutary check on forgetfulness or exaggeration, and more importantly the records form the basis both of my breeding programme and my management decisions. For me, they are an essential management tool. In Dr Miller's words I too can say: "I spend a good deal of time in the house with my record book, studying and planning."[6]

Colony averages

Colony records kept over many years enable me to say something about colony averages, and this may be helpful to other beekeepers, and particularly to those new to the craft. But colony averages are a very personal thing; mine relate only to my bees, in particular locations, managed or mismanaged in a particular way.

Mine, over a period of nearly fifty years, in locations as varied as Devon, Kent and Sussex, Northumberland, and Cambridgeshire and Essex, for eight years or more in each, and in Middlesex and Derbyshire for shorter periods, are perhaps, but perhaps not, more representative than most. They are certainly factual, well documented, and not recollections from memory. Suffice to say that only one year in forty, 1955, was phenomenal, about nine years good and nine near disastrous, and just half the years rather modest. In 1955 my colony average was over 200 lbs. from 24 colonies, winter count, and excluding heather.[7] Only in three other years has the colony average exceeded 100 lbs.

How should one calculate the average crop per colony over the forty or fifty year period? I don't really know. I calculate each year's average on a winter count basis,[8] separately for flower honey and heather honey because not all the colonies are taken to the heather. Even this is not straightforward as I usually carry several 4 and 5 comb nuclei over winter. I count each of these as a colony.

On the basis on the total (flower) honey crop and the total number of colonies put down for winter the average is a little short of 52 lbs. The calculation made twenty-five years ago (for the previous twenty-two years) on the same basis gave an average of 46 lbs. Crops were better in the early years than later, but in the last ten years oil seed rape has restored the overall average to its earlier level.

My colony average (excluding heather honey) for the eight years in Devon was 61 lbs. from around thirty colonies, and varied from 212 lbs. in 1955 to 11 lbs. in 1957. 1959 and 1961 were good years, although not nearly as good as 1955. During the same period the sugar fed averaged 17 lbs. per colony.

The Northumberland average was only 22 lbs. of flower honey over a fourteen year period, but nearly all the forty or so colonies went to the heather, and including heather honey the average was 55 lbs. for the same period. The sugar fed averaged 21 lbs. per colony.

Since I brought the Northumberland colonies down to Essex in 1984, the colony average for the ten years 1985 to 1994 (winter count, about fifty colonies, no move to the heather) has been 76 lbs. Two thirds of this has been from oil seed rape, and about one sixth (in the past five years) from borage. Over the ten years sugar fed averaged 11 lbs. per colony. Except to nucs, we have fed none since 1989.

The heather honey crop has been rather more reliable, and much less variable, than that from flower honey. Over a thirty-five year period (1950 to 1984) from twenty or more colonies (except in the first few years) at the moors each year on moors in Derbyshire, Hampshire, Devon, Somerset, Durham and Northumberland, the colony average was 27 lbs. Only two years (one in Devon and one in Northumberland) were complete failures, and no year really outstanding (as 1947 was). The northern moors yielded a better colony average (33 lbs. over a twenty-three year period) than the southern moors.

Taking bees to the heather involves a great deal of work and considerable cost, and some would consider it not worthwhile. That is not my view, but the beekeeper needs to be fit and active, and keen, and, perhaps, to prefer heather honey to any other, as I do.

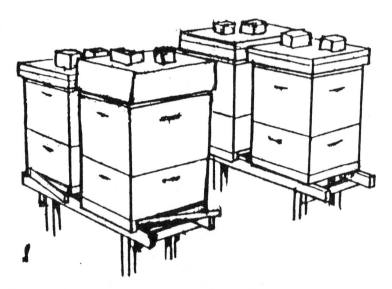

Wintering my 14x12 colonies
14 x 12 above 14 x 8½

6   Colony Management

Management is a matter of getting things done.
The skills that a manager needs to get things done
can only be learned by doing.

Analyse SUCCESS, not mistakes.
From SUCCESS we discover what to do.
From MISTAKES we only discover what NOT to do.

Perceptive observation is a most valuable skill.

Coverdale Management Precepts

H.J. Wadey long ago pointed out that only colonies that reach their prime as the main source of nectar arrives prove really profitable, and that the beekeeper's business is to bring as many as he can to their prime at the right time. But he also pointed out that the happy coincidence of colony prime with the main nectar flow cannot be judged accurately in the fickle climate of the British Isles - no two seasons are alike - and that all management planning needs to be general rather than specific.[1]

Brother Adam's priorities for successful and economically sound beekeeping are a suitable strain of bee, in a hive which imposes no restriction on colony size, and a beekeeper sensitive to the instinctive behaviour and organisation of the colony.

There are obvious and significant differences in climate, flora, and nectar sources between the various parts of the British Isles, for example between the south-west, east anglia, and northumbria, as I and others well know. Brother Adam's priorities are also my own, but the successful beekeeper needs to have a good understanding not only of the bee colony but also of the potential sources of nectar in his locality if anything approaching maximum production is the aim or is to be achieved.[2]

In Kent

Consider, for example, the situation facing beekeepers in the intensive orchard country in the Weald of Kent, where more than half the area within flight range may be in cherry, apple and pear orchards, and nearly half the remainder of the area in woodland of oak and chestnut coppice. The honey flow, if any, is in May, and there is not much to be had thereafter. Moving the bees to other, later, sources, makes evident good sense. In such an area John Hunt and I practised a system of management of our colonies of bees that was really rather complex; more complex than I would care to operate today, but it was well attuned to the local circumstances.

In the autumn of 1954 I wrote the following description of it:[3] "The orchard district of the Weald of Kent offers an intensive spring nectar flow, but little or nothing thereafter. The main, and only too often the only, flow is in May, or maybe in late April and early May; June is usually a period of dearth; and July seldom provides more than maintenance, mainly from bramble, unless the sweet chestnut should yield its dark unpalatable nectar. The season's surplus from colonies remaining in the orchards can usually be assumed to be about equal to the amount in the supers at the end of May.

Early colony development is essential, and to this end colonies are wintered in the orchards in double brood chambers with abundant stores. A few nuclei with young queens are usually carried over winter in single boxes divided into two with a thin wooden partition, but all full colonies are on two boxes, with feed holes and entrances wide open, roofs raised slightly above the cover boards on laths, and excluders as mouse guards between floor board and bottom box. In February or early March floor boards are changed for new ones, excluders and laths removed, and feed holes covered. In early April one or two supers of drawn comb are added above an excluder, and as soon as the cherry blossom is yielding a thorough inspection is made, queens clipped (if not clipped already), and any failing queen replaced by exchanging with one from a nucleus. From then on inspections are made at nine-day intervals or as close to that interval as possible.

Colonies are kept in pairs, and at the next inspection, or sometimes nine days later, one of each pair is removed to a new stand, the one remaining collecting the flying bees from both. Some brood may also be transferred to the moved colonies and replaced with comb or foundation.

In this locality it has been a common experience to find preparations for swarming towards the end of the fruit flow, notwithstanding ample room both for breeding and storing. If one lived on the spot swarming might not be unwelcome at this time of the year, and indeed I do make an artificial or shook swarm if that seems the best thing to do, but the removal of one of the colonies of a pair has greatly reduced swarm preparations in both and has tended to increase the amount of honey in the supers rather than otherwise. This is the time to get foundation drawn, and the colonies remaining on their stands are the ones to do it.

Successful beekeeping depends more than anything upon the strain of bee, and at this time also deliberate queen rearing is started, with the hope, not always realised, of getting young queens mated before the flow from the fruit blossom comes to an end. A breeder queen is purchased from a commercial honey producer from time to time and efforts made to keep the strain as pure as possible. Uniform behaviour of colonies is a great aid to ease of management.

At the end of the fruit flow, or soon after, i.e. as soon as young laying queens are available, the more powerful colonies are split into two and young queens given to the queenless divisions. Supers are left on, so that none may starve in the dearth period.

About 10th to 14th June undivided colonies are moved away from the orchards to the South Downs, there to stay until September, when they are brought back and prepared for winter. Queens heading these colonies are confined to the lower box at the first inspection after moving, or around midsummer day. If queen cells are subsequently found they are destroyed and the queen killed, and at the next visit queen cells are again destroyed and a small piece of comb with eggs and young larvae from a selected queen given in the lower box. Alternatively, one of the young queens raised at home is introduced in a semi-direct cage.

The divided colonies remain at home, and in late July the best of those divisions that retained the old queens are taken to red clover, if, as is usually the case, a seed crop is to be taken from a field not far from the colonies moved earlier. Both lots will be brought back later in one load.

The divisions with the young queens, strengthened as necessary with brood from such others as may still be at home, are taken to heather in early August. In preparing colonies for heather going the dummy boards are removed and twelve combs put into a single box, to ensure both adequate stores and ample brood. Supers are added over an excluder, these being either ekes with starters for cut comb, or drawn comb, often old brood combs, for pressing. On return, the heather and red clover colonies are united under the young queens. Unfortunately, the nearest reliable heather country is 200 miles away, but even in this appalling season of 1954 the trip paid expenses."

Complex perhaps, hard work certainly, but great fun and not unprofitable. We had clear aims and objectives, namely to provide a good and much needed pollination service, to get as much honey as we could in return for our efforts, and to improve our strain of bee. But a similar system of management would have been pointless in Devon and quite impracticable in Northumberland. It illustrates the need to have a system of management suitable to local circumstances, but also illustrates the flexibility that double brood box management can provide, and I shall return to this when writing later about bee management in Northumberland.

In Devon

My beekeeping in Devon was a single handed operation, with bees in two out apiaries and a move to the moors at the end of July. Colonies were wintered on two deep Smith boxes, with an early spring floorboard change, thorough colony check and if necessary re-queening, and subsequent check inspections at close to 9-day intervals.

Management of my colonies in Devon followed what might be described as "standard practice", much as Manley describes in "Honey Farming", and really calls for little comment here except on two matters. First, that I have always added supers well in advance of need for more storage room. The first super should be largely drawn comb (as should the last) and is given initially and at that time principally to provide room for bees, not honey, and avoid any congestion in the brood nest. Second, that with colonies on two deep boxes as a brood chamber the queen should be confined to the bottom box around midsummer day, or perhaps ten days later if the colony is to be moved to the moor for the heather.

The pattern of the season in Devon is pretty much like most of England, I think, although somewhat earlier - a useful spring flow, a bit of a gap, and the prospect, not always realised, of a good summer flow. And there is heather within 30 miles or so, if you choose to move colonies there, although surprisingly few beekeepers do. My colony management there was along lines which can briefly be described as keeping young queens of a good strain heading colonies encouraged to give of their best.

In Northumberland

In Northumberland, as in parts of Scotland, but not, I think, very widely elsewhere, the expected nectar flow from the ling heather in August is the really important crop, and this leads to notable differences in management of bee colonies. My own practice, while not typical, had the same principal aim as that of others, namely the production of heather honey. Flower honey from earlier sources is welcome if it comes, but it is not the principal aim.

I described the manner in which I managed my colonies in Northumberland in an article I wrote in 1970, the substance of which follows. If my bees were still there they would still be managed that way, I feel sure, so the system of management is still relevant. This is what I said:

"Colonies wintered on two boxes, as most of them are, and of good strength in spring, as most of them also are, are split in two about the third week in May and are thereafter treated as two colonies, although they will often be under the same roof. The practice, and the method, is not new, and was developed many years ago by Herman Rauchfuss, an American beekeeper who ran about 1800 colonies for comb honey production in an area where the main honey flow came in August from alfalfa and thus provided a long period of preparation. The Rauchfuss method is described in that excellent little book "Swarm Control Survey" by E.R. Bent, by Wedmore in "A Manual of Beekeeping", and by Pellett in "A Living from Bees".[4] (I describe the procedure, at least as I practised it, in Chapter 9).

Rauchfuss raised his own queens, as I do, but it was Ed. Braun's "Dividing over-wintered colonies for increased honey production", published by the Canadian Experimental Farms Service in 1945, that first aroused my own interest.[5] A somewhat similar practice to the Canadian recommendation of dividing colonies and introducing a purchased queen to the queenless half was in use by Ian Maxwell around Stranraer and Glenluce in the 1960's, and I describe his system later.[6] His aim was the same as mine, namely the production of good colonies for the heather, and I saw some of his divided colonies with fine crops of heather honey in 1965.

My own divided colonies are given second brood boxes and usually make good two deep box colonies, some with supers, by early July. Around 5th July the excluder and supers are removed (separately this time) and set aside, and the upper brood box is also removed and set aside and replaced with an empty box. Into this box all the combs from the upper brood box are shaken, and any considerable patches of drone brood are destroyed with the hive tool. A convenient method of operation is to transfer the combs to a spare box, maintaining the order, as they are cleared of bees. Then remove the box used as a funnel, put the excluder on, and put the box with the cleared combs above it with the supers above that.

Three weeks later, or a little more, all boxes above the excluder are removed, cleared of bees and replaced with two shallow supers with Manley frames fitted with extra thin foundation, with some drawn comb in the lower super. Add a travelling screen, make sure you have an entrance block that fits and will effect closure and push it into the entrance (but not so as to confine the bees) and tap the lock slides firmly into place. Move next morning or on the first convenient morning, at crack of dawn.

The supers, cleared of bees, are later extracted (and are occasionally needed at the moor). The upper brood boxes, largely filled with honey as the brood emerged, are stored away to be returned as second brood boxes to the colonies brought back from the moor in early September. It may be convenient to leave a few colonies behind and stack the brood boxes on these. They can safely be stacked six or seven high if they are securely fixed with lock slides.

My Hexham apiary has more than a hundred massive lime trees within easy reach of the bees. They seldom yield much honey but sometimes they do, and on more than one occasion the colonies left at home and stacked high have got quite a lot of lime honey while the other colonies were at the moor. It is folly to move colonies late in the day or at night if they have nectar to ripen. Early morning is the time to move bees."

Drifting

Research has confirmed what beekeepers have long known, namely that bees may drift within apiaries or groups of colonies to a significant extent, such as to be reflected in the colony yields by differences, due to drifting, of 20 lbs per hive or more.[7] In apiaries where drifting is severe, hive populations become uneven and swarming problems are intensified, with consequent management difficulties and probable reduction in returns. Colony yields become unreliable as an indicator of colony worth (e.g. of the relative value of queens as possible breeders). Disease may also be spread by drifting bees.

The management of an apiary as a unit is greatly assisted if the colonies in the apiary have a similar rate and pattern of development, so that each colony can be treated similarly. Operations to equalise colonies in spring have that objective. The ideal is unlikely to be achieved, but measures to reduce drifting, and perhaps almost to eliminate it, are steps in that direction which present no great difficulty, and ought to be taken.

Three methods of helping bees to identify the hive to which they belong are commonly adopted, namely the use of coloured hives, or of colour at hive entrances; of landmarks such as bushes and trees near hives; and of a layout pattern of hives placing entrances to face different directions. Of course, the three methods are not mutually exclusive, and can be adopted in combination, but the easiest and most practical for the beekeeper, and the most effective, is to adopt a suitable layout.

The layout adopted by Brother Adam at Buckfast is well known. He sets the hives in groups of four, on two stands, with each facing at a right angle to its neighbour, so that if one faces north the other three face east, south and west. The stands are about a metre apart and thus permit the beekeeper to operate on each of the four from the space between the stands. (See illustration at Fig 1).

Other layouts that have been found satisfactory include hives placed in a circle, or in U formation, facing outwards. Such layouts also work well. The presence of a few bushes or small trees will further assist in preventing drifting and will afford some wind shelter.

When I established an out-apiary at Bow in 1956 I adopted a layout similar to that which Brother Adam uses, and since then I have adopted that layout wherever possible. I have found the arrangement extremely convenient and very much to my liking. The groups of four are some 2 or 3 metres apart, to suit the apiary site and my convenience, e.g. to permit passage for a truck or trailer, and to take advantage of bushes or trees, etc. My Tyne valley apiary had stands for 24 colonies, in groups of four, primarily because it is an excellent wintering site. Elsewhere I have had apiaries with three groups of four, with ten or twelve colonies in each, which has suited both my trailer capacity for moving hives and the floral resources of the area.

Colony inspection

Regular colony inspections are an essential feature of colony management. I take the view that a routine of inspection needs to be followed, and therefore needs first to be learnt. Indeed, two aspects of the technique of colony inspection need to be learnt, namely both that of handling bees, keeping the colony under control, finding the queen, finding and destroying queen cells, shaking bees, etc., and also that of handling the hive itself, 1.e. an operational routine.

Not all inspections will follow precisely the same sequence, nor will they all need to be thorough, but a pattern needs to be adopted and followed, so that it becomes routine and habit, thus allowing concentration to be centred on what one is looking for and sees. The purpose of the inspection needs to be clearly in mind before it starts; if there is no clear purpose there should be no inspection. Concentrate on the purpose in mind, e.g. don't look for the queen if you don't need to (although you may see her nevertheless) and if your purpose is to find the queen then look for nothing else, and don't allow yourself to be distracted from that purpose until she is found.

First consider what equipment you may need, what additional boards or boxes, floors or excluders, preferably in consultation with your record book which tells you the state of the colony at the last inspection. Try and anticipate your needs. All but one or two of my colonies are in out-apiaries, fifty miles from home, and there is no going back for equipment, additional boxes, etc. The needs have to be anticipated and provided for.

I always write down, colony by colony, as I go through the record book the evening before the visit, what I think I may need, tot it all up, and collect and load it before I set out. It matters little if one takes too much, but it matters greatly if one does not have what one needs. This preliminary consideration also serves to clarify, for each colony, the purposes of your intended inspection, and this comes back to mind when you open the record book before opening the colony.

Assemble the equipment, boxes, etc., that you think you may need closely convenient to the colonies. Light the smoker and get it going well, so that it will not thereafter go out until you wish it. Have a hive tool, or preferably two, ready to hand. Put on your protective clothing.

Now you can start on the first colony. I generally start with the same colony each time in each apiary, work through that group of four, then on to the next group, and then to the third, starting at the furthest group from the car and finishing at the nearest. Again, it pays to develop a routine. It also helps the memory.

Don't keep a colony open any longer than necessary. Learn to combine speed of operation with perceptive observation. As Wadey said (in respect of demonstrations): "It is one thing to carry out a brief inspection or a needed operation in five minutes, but quite another to keep a hive open for a quarter of an hour or longer to explain ...".[8]

Clipped queens

I have clipped all queens, once mated and laying, for many years, and my method is to clip across both of the larger pair of wings just about the tips of the smaller pair. Queens clipped in this manner can fly a little, sometimes, but never very far, and I have taken and hived more than one swarm with such a clipped queen from low bushes near the hives.

I record in my record book the act of clipping the queen, and subsequently record the fact that the queen heading the colony is clipped. I also record seeing the queen, if I do.

Clipping queens will do nothing to prevent colonies swarming, but it will enable one to extend the period between visits to out apiaries to nine days with certainly that no swarm will be lost.

Foundation

I like to have a good deal of foundation drawn in supers, and if you bring a comb with new honey up from a lower super each time you add a super of foundation bees will draw the foundation into comb pretty readily if there is even a modest income and they are the right strain of bee. But supers added near the end of the honey flow and late in the season (except at the heather) should be drawn comb, if possible, as bees are then quite reluctant to draw foundation into comb. So don't use all your drawn comb in the first supers. New brood combs are almost all drawn from foundation in supers, as I explain elsewhere.

Equalising colonies

I take steps to equalise colonies in spring so that all the colonies in an apiary are more nearly at the same stage of development than they otherwise would be, which both assists in management and delays swarm preparations. I have something to say about this in the Chapter on Swarm Control.

It is an operation, or series of operations, that requires judgement, based on experience, to ensure that the weaker colony gains more than the stronger one loses. It is usually helpful, both to the colonies and to the beekeeper, and can be notably so when all the colonies in the apiary are of the same strain and have queens of the same age.

Migratory beekeeping

Changes in agricultural practice during the past forty years have greatly reduced the sources of nectar available to honeybees in rural areas. Cereal crops are regularly sprayed with selective weedkillers with the result that charlock, once an important source of nectar and pollen, is now seldom seen. The use of applied nitrogen on grassland and the conservation of grass as silage and not as hay, has led to the virtual disappearance of wild white clover as a major nectar source except in upland and hill areas. Forage crops such as sainfoin, which was a prolific source of nectar, and lucerne (alfalfa) are no longer grown.[9]

Herbert Mace could say, as recently as 1952, that "almost any place where chalk comes to the surface is good bee country" and he describes the area around Saffron Walden as specially favoured. Neither is true today.[10] He calls sainfoin "the first of our widely grown fodder crops". I know of none, anywhere in the UK, today.

Instead, we now have an increased acreage of seed crops, notably of oil seed rape, but also of beans and peas, and, as yet experimentally, of lupins, borage, and evening primrose. Of these. not only oil seed rape, but field beans and borage are good bee plants.

Soft fruit, including raspberry, loganberry and gooseberry (mainly on a Pick Your Own basis) and black currant has both increased and is more widespread, and top fruit orchards (mainly apple) remain with us. All these are good for bees. Heather moors also remain, and could profitably support more colonies than they presently do.

There is obvious good sense in moving colonies of bees to oil seed rape, and in so doing beekeepers both equip themselves for, and gain experience of, moving bees to crops where nectar is to be had. Moving bees to crops other than oil seed rape makes equally good sense. I have done so for more than forty years.

Moving bees

When my daughter and I took twenty colonies of bees in my trailer from Devon to Northumberland, more than 400 miles, at heather blossom time, we spent eleven hours travelling, with an overnight stop 100 miles from home and an early start next morning. The bees came to no harm, and did well.

Near the end of July 1975 my wife and I left Foxton at five o'clock in the morning for the 270 mile journey to the Otterburn moors in Northumberland with a load of bees in the trailer. We unloaded and released them seven hours later. It turned out to be a very warm day, and we made several short stops en route to spray water over the screens on the colonies, as my daughter and I had also done. The bees were working the heather and bringing in heather pollen before we left the moor that afternoon, and took no harm whatsoever. In fact these colonies got a good crop of heather honey, as those that my daughter and I took some years earlier had also done.[11]

Properly screened, with entrances blocked, bees travel perfectly well for long distances. They travel best at night, or through the very early hours of the morning, but even through a warm day they come to no harm, provided they can be kept cool and have access to water.

In my view the only satisfactory and safe arrangements for moving colonies of bees over any considerable distance require ventilation to be provided by wire mesh or similar material covering the whole or most of the top of the hive and the entrance closed with a block that permits neither light nor ventilation. In brief a solid and secure entrance block and a purpose made travelling screen.

Though often used, perforated zinc is not an entirely suitable material for a travelling screen. It is much too easily damaged, and the tiny holes are much too quickly closed with propolis by the bees. A heavy gauge one-eighth inch wire mesh is what is needed, and will last for years. It has of course to be framed to fit the outside dimensions of the hive, and mine are all fitted with lock slides so that they can quickly and easily be secured or removed.

110

Hives being moved are often stood upon each other, either temporarily or when travelling in more than one layer. It is vital that the screens do not become totally covered and ventilation excluded, and this can be prevented by adding battens above the screening material. Some screens are so designed as deliberately to permit stacking colonies one upon another without impeding ventilation.

Circumstances often require that the moved colonies retain their screens as cover boards, at least for a time, and on the heather moors quite commonly for the whole time the colonies are there. Top ventilation to the extent that a screen provides would be folly on the moor, when what is needed is to conserve warmth, particularly in the supers where comb is to be built. So screens used for moving to the moors permit a cover board to sit down tight, or a sack to make a tight fit with screen or roof. All my screens permit this, and also permit the entrance blocks to be stored on them, under the roof, when removed.

The form of screen board used by Ernie Pope, which has a wire mesh panel in a cover board and a removeable piece of board to fit the panel, forms an adequate inner cover when the piece cut out is replaced, and also provides space for a folded sack. It is designed for use at the heather, and works well. (See illustration, Fig 14).

All travelling screens will get more or less clogged in due course by the actions of the bees propolising them, and they need cleaning from time to time. Boiling water with washing soda is the stuff to use.

There are numerous ingenious arrangements for securing and quickly closing the hive entrance with a block. For the most part I use standard entrance blocks, solid on three faces, with a 5 inch x 3/8 inch entrance facility on the fourth. A staple driven into each of the side walls of the floorboard prevents the block going in further than it should. The entrance blocks need to be of the right length, a loose fit in the opening, and such as to effect closure when that is intended. A supply of soft string cut into five or six inch lengths will provide one or two pieces to slip round the block to ensure it fits good and tight, and also gives you something to pull on when you need to take it out again. I illustrate this at Fig 22.

Some beekeepers use strips of foam plastic to close hive entrances when moving. I don't like them. Even if satisfactory for their purpose the entrance blocks in normal use have to be removed and carried separately, and subsequently replaced, and there are always a few blocks that don't fit the entrances when they are selected at random. Better use the standard blocks and keep each with the hive to which it belongs. The Askerswell design of floorboard and entrance block (illustrated at Fig 10) is simple and effective, and I have a few of these.

Do ensure that there are no holes or cracks through which the confined bees can get out. If there are the bees will certainly find them. Always carry plasticine in a handy place to stop up any cracks that you have overlooked.

When moving colonies of bees some secure means of fastening floor, boxes and screen together is necessary. For many years I used hive staples, supplemented by metal and later by nylon strapping. Other than picking the hive up again, any operation required the strapping to be cut and hive staples to be removed, and staples and strapping to be refixed for the return journey. The modern hive straps, one for each hive, that can be fastened and released, and refastened again, quickly and easily, and are of sufficient length for a variety of hive sizes, are a great improvement.

Since I have lockslides on all my boxes, and floors, and screens, I need nothing else. The slides can be secured with a small nail, to prevent their inadvertent removal, if need be, but it is not really necessary.

The time to move bees is in the very early morning. Set out at dawn. Then, if anything goes wrong, you can set it right in daylight. I deal with this aspect of moving colonies in the Chapter on Heather Honey.

Wintering bees

The essentials for successful wintering are adequate colony strength, ample stores, a dry hive, and a queen that can be expected to live through the winter and re-build the colony in the spring.

As to adequate colony strength, Brother Adam winters nuclei on Dartmoor on 4 half-size Dadant combs - about the same comb area as 3 BS brood combs. George Jenner did the same. Many others, including myself, regularly winter nuclei of about half single box colony size (five BS brood combs). All these nuclei have young queens of the current year, a preponderance of young bees, and ample stores for their needs. They probably will have both been fed and given store combs from other colonies.

I winter most of my colonies on two Smith brood boxes. Those that have not been taken to the heather get a good and sufficient autumn feed. Those that come back from the moor are almost always nearly solid with heather honey in the single boxes that return, and have added to them the boxes of stores that were removed when they went to the moor (which are often also half full), and get a gallon of feed.

I often take a comb or two of stores from these boxes to make up one or two more to add to other colonies. Two storey colonies winter well with one or two combs less than a full complement in each box, flanked with follower boards, provided they have ample stores.

Don't forget to remove queen excluders at or before autumn feeding time. The queen may have been confined to the lower box in late June or early July by putting the excluder between the upper and lower boxes, and it is easy to forget that the excluder is there. The danger lies in the queen being left below the excluder when the winter cluster moves above it, so remove the excluder and let the queen have access to both boxes.

Sometimes colonies returning from the moor have to winter in their single boxes, without feeding, as they did in my Tyne valley apiary (unavoidably) through the winter of 1975/76. I lost only one in a group of twenty-six. I gave them all a good spring feed, gave them second brood boxes of empty combs and a second feed at the next visit, and they turned out thoroughly well.

Almost all winter losses are due either to starvation or to the colony losing its queen. Neither is wholly preventable, but in both respects it pays to do one's best, both by ensuring an adequate store of food and by replacing queens before their third winter.

I don't rely on queens that have headed strong colonies through two full seasons to survive a third winter or to be satisfactory in re-building the colony in the spring if they do.

Some will, and some will be superseded by an early raised daughter queen, but I don't rely upon supersedure for my queens.[12] If you regularly re-queen colonies so that very few queens are expected to head full colonies after their second full season you won't have many winter losses from queenlessness. My chosen breeder queens are taken through their third winter in a single box or in a nucleus, and thereafter kept in a nucleus.

A reserve of food combs is very useful, and I usually winter some colonies on three deep boxes so that I can draw food combs from the top box (an added box of stores) to give to other colonies if the need arises. Put an excluder beneath the top box at the earliest spring visit if you don't want to find brood there. If need be, also, starvation can be averted by giving candy or fondant, even moistened bags of sugar, from December onwards. I refer to this in the chapter on Feeding.

A dry hive is another essential, and a good wintering site is a great help in this, and in other ways. Try and find a site that has a maximum of winter sunshine, which both helps to keep the hive dry and to provide maximum opportunity for winter cleansing flights. Shelter on the sunless side helps to create a winter sun pocket. If, in addition, the site has early spring pollen within easy reach of the bees it can be near ideal. My Tyne valley site had all this. Regrettably there is generally little there for bees after the sycamore in late May.

Hives in pairs on stands, kept clear of too much grass or weeds, will have good air circulation around the hives. I have entrances full open (the full width and depth that the floor provides) with mouseguards. A queen excluder between floorboard and brood box is an excellent mouseguard, but it is not always convenient to insert it, and it must be removed in early spring.

I have feed holes in cover boards open and uncovered, and tile battens (of about 1¼ x ¼ inch material) to raise the roof above the cover board. Colonies winter well with the cover boards removed and replaced by two inch deep rims on which tile battens support the roof.

Colony management for honey production.

For most beekeepers the production of honey is the principal aim. Colony management is a vital ingredient, and because of its importance, I set out below, in summary and as briefly as I can but nevertheless at some length, what I have tried to say in more detail on this vital subject elsewhere in this book.[13]

The single most important factor in maximising yields is the strain of bee. Colony management is vastly easier, and far more rewarding, with a really good strain of bee, and with the same good strain of bee in each and every colony.

Seek out really good bees, and keep the best strain you can find. Eliminate worthless bees, and bees with undesirable characteristics. Give the highest priority to reluctance to swarm. Take positive steps to ensure that the colonies are headed by young queens. Re-queen production colonies every second year.

Learn how to handle bees - with a minimum of smoke but always under control; carefully, but speedily and firmly; and consciously try to improve your observation and understanding of the state of the colony. Never open a colony without a clear purpose in mind.

Keep colony records. I keep mine in a record book, and use a simple shorthand. Each colony is tagged with a number, and each two facing pages on opening the record book records a single colony. For me, my record book is an essential management tool. A glance at the colony history before opening the colony is an essential first step; a note before moving to the next colony is a long established habit. Study of the record book before making the visit ensures that I take with me to the out-apiaries what I am likely to need there. Careful study of the colony records is the basis of my choice of breeder queens.

To be effective, the management of colonies must be based on the circumstances of the individual beekeeper. For each of us these will differ. My situation with all my colonies in out-apiaries distant from home, except for two or three used for queen rearing, will not be that of many other beekeepers. For the beekeeper with two colonies in a town or suburban garden, with neighbours to consider, priority must lie in keeping good tempered bees and in ensuring that the colonies don't swarm; and the frequency and the timing of colony inspections has to have the neighbours in mind.

Appropriate and effective colony management depends on the locality, the beekeeper, the bees, the equipment, the number of colonies, whether they are in out-apiaries or at home, and the weather; which differ for each of us (except perhaps for the weather). The management system should reflect both the probable nectar flows, and the amount of time that one is able to devote to beekeeping.

An essential aim of spring management is to produce the maximum number of foraging bees at the time the main honey flow starts. To achieve this, the brood peak has to be reached five to six weeks before the flow, and we need to ensure that no food shortage hinders development for some two months before brood peak. For a main honey flow in late June, or later, it can be done. For oil seed rape, the brood peak needs to be reached before the end of March, and that is just not possible. The best we can do is to encourage early development and get somewhere near it.

You cannot have brood maximum, which you want some time before the main honey flow, both in late April and in early June. Colonies that reach their peak late in an early district bring no profit; and colonies that are very strong very early in a late district will probably swarm, and even if they do not they will be past their best when the main honey flow comes. Try to work to the date that best suits your district. Success (at best only partial) is one of the tests of good management.

Keep colonies as strong as possible at all times. We cannot be sure when the honey flow will come, nor know which are going to be the hard winters. In swarm control, aim to keep the hives as full of bees as you can. Spring division controls swarming and gives increase without much loss of harvest; the total crop will often be more, not less. But don't split colonies after the end of May. What we must not end up with is weak colonies for the winter.

My own methods

I use Smith hives, and work the colonies on double brood chambers; on two deep boxes, National size, each holding eleven 14 x 8½ inch combs and a dummy, throughout the year. National hives, if I used them, would have precisely the same capacity. I much prefer Smiths.

116

With oil seed rape everywhere, and no clover flow, with the main source of nectar not in July but in May, we need an early brood peak. We had the same need years ago in the cherry and apple orchards in Kent where I kept bees. The only way I know to achieve it is to winter really strong colonies with young queens and ample food and encourage their development in spring.

My colonies are wintered on two deep boxes. From February onwards I give the colonies pollen patties, directly on the top bars. In early to mid March I reverse the two boxes. I provide a clean floor board, set the upper box (in which all the brood will be at that time) on the floor board, and set the former lower box on that; give a second pollen patty on the top bars and cover. In early April each colony is given a gallon of sugar syrup. A Rowse Miller box feeder serves as a cover board on every colony, is turned upside down to cover the pollen patty, and turned again to receive the syrup.

I make my first inspection of colonies towards the end of March or during the first days of April, and try and choose a warm spring day with little or no wind and some sunshine. This is the time to assess colonies for strength, for stores and for brood, and for brood disease.

I don't particularly look for the queen at this time, but if I see her I check that she is clipped and clip her if she is not. It is at this time, or at the next inspection, that I clip the young queens that were reared and mated the year before, and I try and have all the queens clipped by the end of April - by the time the rape flow starts.

This is the time to start equalising colonies. In each apiary there will probably be one or two that are weaker than the others, and after checking them all for stores, brood and disease, I swap the weakest with the strongest.

I take a weak one and put it in the place of a strong one and move the strong one to the weak one's place. I try to achieve nearly equal colonies in each apiary by the end of April. They will diverge later, but management is vastly easier, and swarming problems greatly reduced, if one can achieve nearly equal colonies by oil seed rape time.

On my second visit, later in April, I again take steps to equalise the colonies, this time by exchanging a frame of brood and bees from a strong colony with a relatively empty comb from a weaker one, or by shaking bees off a brood comb from one colony into another. It is as important to reduce the strength of the strongest colonies at this time as it is to assist the build up of the weaker ones.

By late April the colonies will need more room, and I put a queen excluder and a super of drawn comb on every colony a week or two before the rape flow starts. I give the first supers before they are needed for incoming nectar. They are given to accommodate bees, whether or not the nectar flow is imminent. Overcrowding in the brood chamber at this time will lead to swarming later, and must be avoided. Many beekeepers super far too late and create congestion, which forces swarming on colonies that might otherwise have gone through the season with no trouble.

In late April/early May, i.e. as soon as the rape flow starts, I give a super of foundation beneath the first super - a deep box of foundation to get new brood combs drawn to the extent that I require them - and thereafter add supers on top as required. When the rape flow comes to an end I take off all the supers, and use fume boards with Benzaldehyde or Canadian type clearer boards to clear the bees from the supers to load and take them home at once and not a day or two later. I extract the honey from them at once also. As soon as convenient thereafter I give a gallon of syrup to each colony.

Swarming

Swarming at the rape is a problem for all of us, and shortage of room to store and process nectar is a potent cause. A hot spell will cause nectar to come in so fast that all the available storage room is used up very quickly. It is the very best time to get new combs drawn from foundation, but foundation doesn't provide room for nectar storage until it is drawn out, and I try and keep ahead of the bees with their need for storage space at this time.

Some colonies will build queen cells at the rape, no matter what we do. I may kill queen cells once, but I always split the colony on the second occasion and usually shake a swarm. I find and take the queen and put her, with the bees shaken from six combs, into a deep box of brood combs, and take

118

the shaken swarm to a new site. Or I use a Taranov board, which allows me to leave the shook swarm in the apiary. Nine or ten days later I check for queen cells in the colony from which the shaken swarm was taken and destroy all but one, or destroy all the queen cells and introduce a virgin queen from a selected breeder.

Once this first swarming phase is over, the bees will usually settle down and build up gradually towards the summer honey flow, whenever that may come. Colonies on a midsummer crop such as field beans need watching, and will need room, if there is a flow, when colonies elsewhere are barely holding their own. Midsummer swarming is the ruination of the honey crop, and must be controlled.

I allow all colonies (whether swarmed or not) the run of two brood chambers (adding supers as necessary) until midsummer. Three weeks through June I put a queen excluder between the two brood boxes after smoking very generously from above. At the next inspection, I look for eggs in this box. If none are present the queen is evidently below, which is what is desired, and what usually happens. If eggs are seen in the upper box the queen is upstairs, and either I find her and put her below or I transpose the boxes. I remove the queen excluder when the supers are removed at the end of the season and feed for winter.

Double brood chambers

Advocating use of the 14 x 12 inch frame in a 21-frame hive Robin Dartington writes: "Few beekeepers would deny that any brood nest in two boxes is not easy to manage. The bees fiercely resent the boxes being separated, and the space between the two sets of combs is repeatedly bridged with brace comb, since the bees find a gap within the brood nest unnatural."

I don't agree with that at all. It is not my experience, which is considerable. The use of two or more boxes for the brood nest is a universal practice, world wide, and it just wouldn't be so if it made management more difficult and the separation of the boxes was fiercely resented by the bees.

I don't find the bees resent the boxes being separated, or that the space between the two sets of combs is repeatedly bridged with brace comb. It depends to some degree on the bees;

not so much on their temper as on their tendency to build brace comb - some will build brace comb everywhere - some build very little anywhere.

But it depends more on the maintenance of the correct space between the two sets of combs, and on the design of the frames. Wide top bars, and full width bottom bars, inhibit brace comb. The depth of boxes must be such that there is a bee space between them. Boxes tend to shrink and become less deep than they need to be, and it is when this happens that the two sets of combs are stuck together.

Regular removal and replacement of combs at routine inspections of the upper box will tell you whether the bee space is right, and the occasional scrape with the hive tool will keep the bottom and top bars clean. Add a thin fillet all round to put right the boxes that shrink.

To keep bees on a deep and a shallow, as many do, is to adopt a double brood chamber system. It requires the same attention to maintenance of the bee space between the two boxes, and to elimination of brace comb in that space. I consider it better to use two deep boxes. The facility to exchange boxes, for replacement queens to mate and build up colonies in a box above the supers, for interchange of combs between one box and the other, are invaluable assets, that are not present in the deep and shallow system.

In my experience the use of a single box of 14x8½ inch frames as the brood nest will inevitably lead to the storage of pollen, often in large quantity, in the first super, which in turn leads to that super becoming part of the brood nest, and thus to the adoption of the deep and shallow system. I hate pollen in supers, and the only way to avoid it that I know is to give the bees sufficient room - and that means sufficient depth - below the queen excluder in one box or in two.

Although I carry a number of single box colonies over winter - strong nucs with summer mated queens in single Smith boxes - I don`t run any of these as single box colonies in the following year. The queens would find the brood space inadequate and the bees would use the first super for pollen storage. So they get a second brood box as soon as they need it.

120

My Hexham apiary, showing arrangement in groups of four

Damage by deer
But note how lockslides saved the day

BS frames newly assembled and fitted with foundation

My uncapping and draining tank
It holds 22 combs

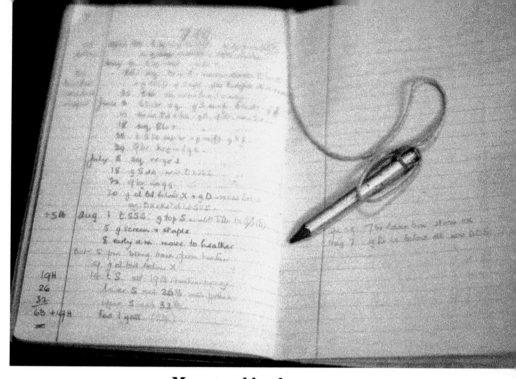

My record book
A year's record for each colony at each opening

My bee box and contents

Taking a swarm with a clipped queen

**Upper brood box set back to provide an upper entrance
as Wadey did routinely**
Note also the lockslides

14x12 box of foundation added for new comb to be drawn

One of my apiaries in Essex with pollen trap in use
Note 14x12 brood boxes

Another of my apiaries in Essex, showing hive number tags

A 20-acre (8 ha) field of borage in Essex

Start of woodpecker damage - at handholds and between boxes

Woodpecker damage. A hole made in July 1995
Note earlier damage at handholds and sticky tape cover

My Hexham apiary with my car and trailer in 1983
Note: can get trailer close up to hives

Wild comb built in a super that had no frames

128

7   Management systems

Any beekeeper who has a long experience and has moved about a little is bound to wonder whether the system he has built up for himself would not have been very different if he had lived in other districts..

K. K. Clark, "Beekeeping" (1951)

While I don't suggest that one should adopt any system of colony management other than tentatively, I do think it sensible to study systems which other beekeepers have developed and found satisfactory; and when the system seems to be suitable to your own circumstances and needs to try it when opportunity arises. You may find a system that suits you very well.

I set out below two systems of management developed by other beekeepers that I have found very satisfactory and have frequently used, and also describe two systems used by others in heather districts. I add one of my own.

Wilson's system

A system of management in which young queens are reared and mated from nuclei under the same roof as the colony was developed by the late R.W. Wilson, who gave a very full outline of the method in the May 1948 issue of "Beekeeping" (the journal of the Devon BKA).[1]

I used it in that year on two colonies that I had near Lewes, in Sussex, and found it simple and effective, and very suitable for use on colonies in an out apiary, as these were. I have since used Wilson's method, with or without slight modifications, many times, on colonies in Kent and Sussex, in Devon and in Northumberland.

Wilson lived and had his bees at Barrasford, north of Hexham, in Northumberland, and I later met many beekeepers who knew him and one who worked with him.

Wilson's method was one of three selected for investigation by the BBKA Research Committee and reported upon by Wedmore in 1952.[2] He then said: "Briefly, the method provides for queen raising, and for increase if desired, as well as a method of swarm control. There is no need to find the queen, at least until she is to be superseded. Drones are not confined at any stage, and the management of nuclei is simplified." Together with A.W. Worth's "Worthwhile Method", Wedmore liked and recommended Wilson's method.[3] I now have a long experience of it, and so do I.

The nuclei should be made as early as possible, about the time that the first drones are flying and before preparations for swarming might be expected. The date will obviously vary with the district and with the season.

A cover board is required in which entrances have been cut in the upper rim on the two sides (not the front and back) about an inch and a half wide, or pieces cut out and replaced with a nail through the middle to swivel. A division board will also be required which will make a bee-tight fit with this cover board when placed in a brood box above it, to divide the box into two equal parts, each with an entrance. The cover board may provide a solid floor or may preferably have two pieces cut out and covered both sides with wire gauze, one each side of the division. A suitable board can easily be made from a clearer board.

On a suitable day transfer to an empty brood box two good food combs, and between them two combs of young unsealed brood and eggs and two combs of sealed and emerging brood, the combs with eggs and young brood being in the middle. Brush or shake the bees off these combs when transferring them, thus making sure that the queen is not taken into the new box. Close up the combs remaining in the colony and put spare drawn combs or frames with foundation on the flanks to fill the box. Put an excluder on top, and then the new box with the brood and food combs above that and replace the cover and roof. Leave for about two hours.

All that is then required is to reverse the positions of the colony and the new box. To do this, lift off and set aside the top box. The combs will now be well covered with bees. Remove and set aside the colony, leaving its floor board or providing another in the same place. Set the new box on the floor board. On top place the screen board in which entrances have been cut in the upper rim, and see that the entrances are open. On this screen board put the original colony (without it's floor board, of course). It matters not whether the colony was originally in a single box or in two. If it was in two, put the two boxes above the screen board in their original order, i.e. bottom box still at the bottom, and treat the two boxes as one brood chamber throughout.

The flying bees will now enter the bottom box, which is queenless, but has eggs, young and older brood and nurse bees. Queen cells will soon be started. The top box (or boxes) will contain most of the brood, plenty of nurse bees, and the queen. Within a day or two bees will be using both side entrances.

Ten or twelve days later reverse the boxes once more, the broodchamber with the queen on the floorboard, and the box with the queen cells above the screen board. First check for the presence of the queen by inspecting to find young brood. Then set the brood chamber to one side, without the screen board. Check to see whether the queen is on the screen board, or preferably shake the bees off the screen board into the moved brood chamber. Set the bottom box, now with queen cells, to one side, without the floor board. Set the brood chamber, with the queen, on the floorboard, add an excluder and a super, then the screen board, with entrances open, and finally what was the bottom box, with queen cells.

Now divide the top box into two by inserting the division board, first ensuring that there are queen cells in both halves. There is no need to select or destroy cells.

There will now be two nuclei each with a queen cell or cells and an entrance which bees have become accustomed to use. In due course the young queens will emerge and mate and there will be three colonies under one roof.

The nuclei can be used for increase, e.g. to build up for the heather. One of the nuclei can remain and allowed the full use of the box, and later be united to the original colony. If one queen fails to mate it is easy to unite the two nuclei - merely replace the division board with a dummy board i.e. a board of the same size as a brood frame, and later take the dummy board to one side.

If increase is not required, or no more than one new queen to be raised and mated, there is no need to divide the top box. The screen board can then be simpler and only one entrance (which can then be at the front or back, or at one or the other side) is needed into the top box.

If it is desired to raise all the queens from a selected breeder queen the simplest way is to put a comb or two with eggs and young larvae taken from the breeder queen in place of the two combs of young brood when the first operation is carried out. Spare cells raised on these combs can then be exchanged later for cells raised in other nuclei. Wilson destroyed the first cells raised and then gave queen rearing material from his breeder queen.

Wedmore suggested that A.W. Worth's "Worthwhile Method" could usefully be employed at the time of putting the nuclei up and dividing the box. This requires action as follows: Before reversing the boxes at this time find the queen, put her with the comb and bees in a spare brood box, put a comb of brood and bees on each side of this comb and fill this box with good empty drawn brood combs (say two on each side) and foundation. Close together the combs remaining in the box from which the queen and combs of brood have been taken and fill the flanks with empty comb and foundation. Put the new brood box with queen and brood in place of the super, i.e. above an excluder and below the screen board. Remove the excluder about eighteen days later (not less than 14 days nor more than 21).

When I tried this I found, as Wedmore suggested, that no queen cells were raised below the excluder and the whole unit could safely be left for three weeks. I have since used the system many times with the same result, sometimes on all the colonies in an out apiary, which can then be omitted from routine visits for a time. But don't overlook the possibility of supers being required in the meantime.

Diagram 4   Wilson Management System

1. Divide over-wintered DD colony, AB, in spring, rearrange, and g X between boxes, thus:

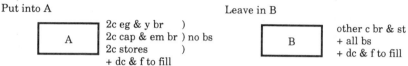

Put into A

A	2c eg & y br)
	2c cap & em br) no bs
	2c stores)
	+ dc & f to fill

Leave in B

B	other c br & st
	+ all bs
	+ dc & f to fill

2. g X and leave for 2 hours,

3. after 2 hours, exchange boxes, and replace X with board with entrance each side, thus:

A w no q & no bs

A
X
B

B w q & bs

split board

B
=
A

w q

w entr each side

w no q

4. After 10 or 12 days, when qc built in A, again reverse boxes, g XS to B, and divide A,   thus:

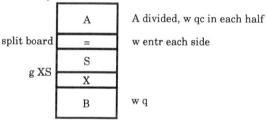

A	A divided, w qc in each half
=	w entr each side
S	
X	
B	w q

split board

g XS

OR, 4A, plus Worth method, t q w 3c br into new box C, g dc & f to fill, and g XC to B instead of XS as above, thus:

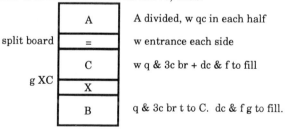

A	A divided, w qc in each half
=	w entrance each side
C	w q & 3c br + dc & f to fill
X	
B	q & 3c br t to C. dc & f g to fill.

split board

g XC

and, 14 to 21 days later, remove X.

Wilson used a hive of his own design, based on a short lugged hive long in use in Northumberland, and essentially a ten comb Smith. He wintered his colonies on two of these boxes, and took most or all of them to the heather. His method of management is equally applicable whether the colony is initially in one or two boxes, but a single box colony needs to be pretty strong in brood and bees at the time of the initial operation.

John Ashton developed a modified Snelgrove system for heather going beekeepers in Northumberland which worked well.[4] I found Wilson's system simpler and equally effective, with the great advantage of being entirely suitable for use in out apiaries.

Spring Division (Rauchfuss Method)

The Rauchfuss system of management is essentially one of spring division, and was devised and adopted in an area of USA where the principal flow was in August, from alfalfa (lucerne). I found it highly appropriate to the very different climate and flora of Northumberland but the essential similarity that the main flow is in August, from the heather. I used it regularly on my colonies in the Tyne valley, with no problems and considerable success. As Wedmore says, spring division gives increase without much loss of harvest, if the colony is fully developed well in advance of the main flow.[5]

The system is very simple. Colonies are wintered on two brood boxes amply supplied with stores, sufficient to carry them through the uncertain spring period without qualms. Such colonies can be expected to be in really good fettle, with brood and honey in both boxes and in need of a super, around the third week in May. At that time the upper box is lifted off and set aside, an excluder put on the lower box, then a comb honey super. On top of this super put a clearer board such as that normally used with Porter escapes, with an entrance cut in the rim at one side or the back to give access to a box above. This entrance should be open. Cover the escape hole or holes (mine have two) with a piece of excluder zinc. Replace on top of this board the upper box that was set aside. The bees have access to and from the upper box through the small pieces of excluder zinc, which tend to inhibit free movement. The upper entrance is soon in use.

Eight or nine days later, examine to see in which box the queen is laying, and take that box, as found, to a new site to build up as a new colony.

Check that there are queen cells in the remaining box and leave this box on the old site on a floor board with the excluder and super above. A new floor board, cover board and roof will be required.

Diagram 5   **Rauchfuss Method**

1. In May, divide DD colony, AB, without looking for q,, & g XS, thus:

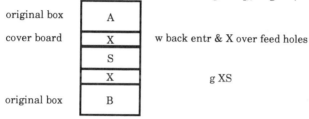

2. Eight or Nine days later, t D w q (A or B) to new site. Leave D w no q (B or A), now w qc, on old site, thus:

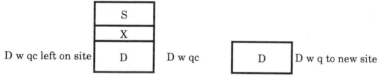

OR, leave both A & B on site , thus:

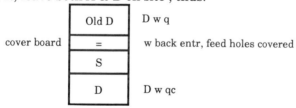

Rauchfuss destroyed the queen cells and gave a newly emerged virgin queen to the extent that he had such queens available, but otherwise made use of the queen cells raised in the queenless portion. A young mated queen can be given if one is available. I have usually made use of the queen cells, but used selected cells from a colony from which breeding was acceptable.

Instead of taking the box with the queen away to a new site, I have often varied the procedure by leaving the box with the queen on top of the clearer board, with a back entrance, but under the same roof, moving it away later, or uniting with the other box under the new queen. Both portions usually build up to make good units for the heather, so uniting is only a fall back for failure in one of the queens.

The system is so simple, and so effective, that I came to use it on all colonies in my Tyne valley apiary that I judged to be of sufficient strength at the mid May inspection. Aimed at an August honey flow, from the heather, it proved very suitable. It might not be so in other circumstances and situations.

Whether or not the system as a whole is suitable to one's situation and circumstances, the initial operations provide a simple and satisfactory method of getting queen cells built in a selected colony without having to find the queen, and thus could be more widely adopted with advantage.

Ian Maxwell's method.[6]

Maxwell's method is very simple. The colony is wintered in a double brood chamber, and in late spring, when both boxes contain brood, all the bees are shaken from the upper box into the lower box, and the upper box replaced above an excluder with the combs in their original order. Bees soon come up through the excluder to care for the brood in the upper box and to occupy the box once more.

Next day the excluder is replaced by a split board [7]with an entrance at the rear, and a new queen given in a Butler cage in the top box. From then on, when the new queen has been accepted, (as she nearly always is), the two boxes are treated as two colonies and built up for the heather.

Diagram 6   Ian Maxwell's Method

1. In late spring, divide over-wintered DD colony, AB, thus:

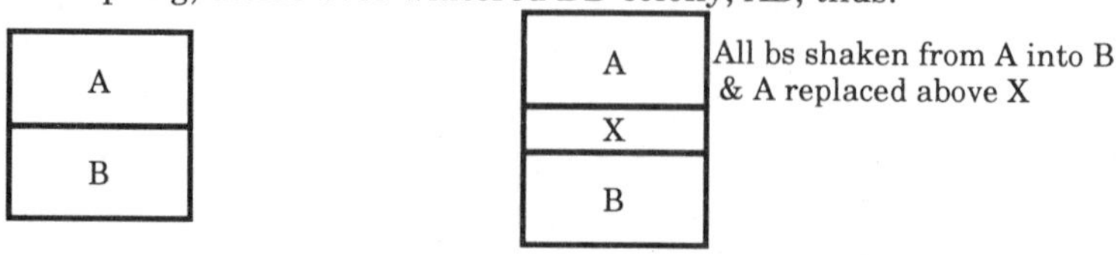

```
+---------+        +---------+
|         |        |    A    |
|    A    |        +---------+
|         |        |    X    |  All bs shaken from A into B
+---------+        +---------+  & A replaced above X
|         |        |         |
|    B    |        |    B    |
|         |        |         |
+---------+        +---------+
```

2. Next day, replace X by a split board w back entrance, and g new q in Butler cage to A. Leave thus:

```
┌─────────────┐
│      A      │   new q in Butler cage g to A
├─────────────┤
│      =      │   split bd, back entr
├─────────────┤
│      B      │   o'w q in B & all flying bs.
└─────────────┘
```

Bert Mason's system.[8]

Thirty miles south of Aberdeen, Bert Mason winters his colonies in single Smith brood boxes. Towards the end of March, or early in April, he gives all the overwintered colonies a gallon of syrup, in a Rowse Miller feeder. The first inspection of colonies is made in late April or early May, when the queens are clipped and marked, if not so already, and the colonies equalised by shaking young bees from strong colonies (one or more combs) into weaker ones, to level them up. A few days later an excluder and super of drawn comb is added to each colony, and all are moved to oil seed rape.

In the last week of May or the first week of June (essentially before they would otherwise swarm) all the colonies (except those considered insufficiently strong) are artificially swarmed. From each colony the queen and the bees from six combs are shaken into a new box of drawn comb, which is then covered and closed and moved to a new site. A brief check to ensure that the queen is laying, and to add excluder and supers as necessary, is all that the moved colonies (the artificial swarms) require.

Ten days after the artificial swarms are made the colonies from which the swarms were taken are checked for queen cells and queen cells reduced to one. A subsequent inspection is made to check that the resultant queen is mated and laying, and supers are added as necessary.

In mid July the queens are killed in those overwintered colonies that were not artificially swarmed, and either the colonies allowed to re-queen themselves, or queen cells from a selected source are introduced.

Flower honey, from oil seed rape, from raspberry, and from bell heather, is removed and extracted, and all colonies are moved to the ling heather moors with a super of drawn comb given the day before the move, which is made at dawn. Heather honey supers are taken off the colonies at the moor, Rowse

Miller feeders added, and two gallons of syrup given to each colony. The colonies are subsequently brought back from the moor to their wintering sites when convenient. The Rowse Miller feeders are generally left on the colonies over winter.

To the extent that new drawn comb is required, foundation is given in the supers both at the oil seed rape and at the raspberry, but not at the heather. New brood combs are drawn in supers at the oil seed rape.

Diagram 7   **Bert Mason's Management System**

1. A — Overwintered in single box

2. S / X / A — g XS & move to osr

3. In late May or early June, art sw all colonies except those insufficiently strong:

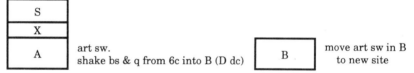

S / X / A — art sw. shake bs & q from 6c into B (D dc)

B — move art sw in B to new site

4. Ten days later, reduce qc in A to one only, that new q in A is laying. g SS as necessary.

5. Check, in due course,

6. In mid July, re-queen all colonies not art sw.

My system of management with colonies on 14 x 12 frames[9]

I use the 14 x 12 inch frame in a management system that is an adaptation of that used by Raymond Zimmer in Alsace. Zimmer uses 12-frame Dadant and 12-frame Langstroth hives and Buckfast bees in an area where oil seed rape in spring and spruce in late summer are the principal sources of nectar. His crops are legendary. I use Smith hives that take 14 x 12 and 14 x 8½ inch combs. My colonies winter on a double brood box, 14 x 12 above 14 x 8½ (Zimmer's on Dadant above Langstroth) and are run from March to September on a single

14 x 12 box. In an oil seed rape area with a later flow in most years the system has proved very successful and very much to my liking. I have increased the number of 14 x 12 colonies managed in this way each year, and now have twenty-four.

From early March to mid September the colonies are on deep Smith boxes with eleven 14 x 12 frames and a dummy, plus a queen excluder and supers as required. I start rearing queens on the oil seed rape flow, and aim to have good single box (14 x 8½) colonies headed by these queens by late summer. In mid September I find and remove the queen from each of the 14 x 12 colonies, and either kill the queen or cage her with attendant bees for use elsewhere. To each colony I then add, above a sheet of newspaper, one of the single box colonies with a young queen. Ten days or so later, by which time the two colonies will have peaceably united, the two boxes are reversed, putting the 14 x 8½ box down and the 14 x 12 box on top. The colony winters on the two boxes.

In early March I remove the lower box and leave the colony with just the one 14 x 12 frame box, into which I shake all the bees from the box removed. I usually replace the removed box, after shaking the bees from it, on a queen excluder above the 14 x 12 box. The boxes removed at this operation usually contain unused winter stores and need bee proof storage until they are once more put into use. Replacing them on the colonies above excluders meets this need and also ensures that the colonies are not denuded of stores. They are very unlikely to have brood in the combs (never, in my experience). The exchange of boxes after uniting is intended to ensure that in early March the brood is all in the 14 x 12 combs, so that the 14 x 8½ box can be taken away, and it does just that, but in the event of one of the removed boxes having brood in one or more of the combs, back on the colony above an excluder is the only sensible place to put it.

Diagram 8   My system with 14 x 12 frames

1. In mid September kq in 14 x 12. Add above newspaper 14x8½ colony w yq
2. Ten days later reverse the boxes, thus.

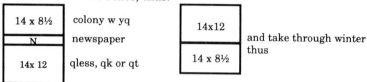

3. In mid March take away the 14x8½ box, clear it of bees, and put it back on the 14x12 box above an excluder, thus :

4. Late April onwards, take the 14x8½ box for queen rearing or swarm control. Add supers to 14x12 as required, and build up a 14x8½ colony for autumn uniting.

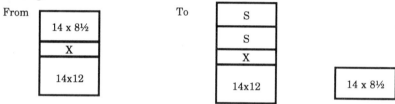

The boxes with 14 x 8½ frames removed at this operation will be required by late April/ early May for swarm control and queen rearing purposes. They will be the boxes into which I shake swarms, as a method of swarm control, and the boxes in which nuclei with newly mated queens build up colonies for the autumn uniting. So that these boxes may be readily taken when they are needed, I keep them on top, and put supers under them as soon as the rape comes into flower. To the extent that I need new brood combs, the supers given are deep boxes, of either size, with foundation.

We no longer make routine checks for queen cells in the 14 x 12 colonies during the rape flow; they are all headed by queens in their first production year, of a strain of bee reluctant to swarm, in a brood chamber big enough not to restrict the queen. We never found cells when we did check, so we stopped checking. It saves a lot of time.

8   Swarm Control

Modern beekeeping, at least in English speaking countries, demands above all a bee that is reluctant to swarm.

Brother Adam, "In Search of the Best Strains of Bees".[1]

Necessity and means

Every colony is a potential swarmer at least once every year. The construction of queen cells is the first indication that the colony may swarm. Queen cell cups, along the bottom of combs, can be ignored. But when larvae are present in queen cells, and are being fed, the probability is that the colony will swarm unless the beekeeper takes steps to prevent it. A probability, not a certainty, as until queen cells are well advanced, perhaps near sealing, changed conditions, such as adverse weather or a sudden loss of bees, may cause them to be aborted.

It is common practice to destroy queen cells, when found, if none have yet reached the point of sealing. Sometimes no others are built, but more often there are fresh queen cells at the next inspection, and appropriate steps then need to be taken to satisfy the swarming urge or to see that a swarm is not lost. Much depends upon the strain of bee. Some are inveterate swarmers.[2]

I doubt the existence of a strain of bee that never swarms, but experience soon teaches that some strains swarm immoderately, some regularly, and some rarely. I do my best to keep bees that are disinclined to swarm; I keep very few queens beyond their second year; and I give colonies room in advance of requirements. Each of these helps to reduce swarming; together they go far to eliminate it. Even so, some colonies will build queen cells, in apparent preparation for swarming, and it is best to assume that they will swarm unless appropriate action is taken.

When queen cells are first found, and destroyed, it is worth while adding a box of empty combs, if such are available and the colony can be encouraged to use them. Overcrowding is commonly a prime cause of preparations to swarm.

If the colony has honey in the supers and the flow is still in progress, one way of alleviating overcrowding is to extract the honey in the supers and return them to the colony. I am told that it is very effective, and I know one beekeeper who has and uses a small hand extractor for just this purpose. Extracting honey, other than at the end of the season, is an operation best avoided, in my view, unless there is no alternative; evidently some think differently.

Swarm control is essential to successful beekeeping. Of that there really is no doubt. The beekeeper who can be sure of being there when one of his colonies swarms must be rare, and even for him a swarm may alarm his neighbours, and cause considerable nuisance. For him, no less than for the commercial man, controlling swarming will increase his crop, particularly in poor years, if it is well done.

There is no doubt about the necessity for control; it is the means of control that are frequently not well understood, and give rise to most argument. The beekeeper needs to learn to work with the colony and not against it, and to avoid operations and methods that frustrate the colony urge. That requires understanding and perceptive observation of colony behaviour, and at least a little knowledge of what best to do in a variety of circumstances.

I, and I suppose most beekeepers, much dislike those strains of bee that swarm and swarm again - that typically throw a smallish first swarm and a cast with every virgin - and there are such. My strong preference is for a strain that throws a large first swarm - if it swarms at all - and then settles down under a new queen without casts - and again there are such. The first step in effective swarm control is the acquisition and

142

maintenance of a good strain of bee. The second is so to manage the colonies as to minimise their natural tendency to swarm.

I like to visit out apiaries in late February to give pollen patties, and see that all is well, note the extent of woodpecker damage, if any, and, if bees are flying, to make a first estimate of colony strength. I make a second visit in mid to late March, and choose a good flying day, if I can. I then give clean floorboards, renew the pollen patties, and, if there are one or two colonies much above average strength, and one or two relatively weak ones, as there usually are, I exchange one for the other, moving one bodily to the other's stand, and vice versa. With the arrangement of colonies in the apiary that I use to minimise drifting (groups of four) it is usually only a matter of exchanging one of the four for another in the same group. This is the first step towards equalising the colonies.

I make the first colony inspection, a thorough one, in mid April, here in East Anglia. It used to be a month later in Northumberland. Again, I take steps to equalise the colonies, and shake young bees from strong colonies into weaker ones, or take brood combs and adhering bees from strong colonies and give them to weaker ones, (first making sure that the queen is not on the combs taken), in exchange for combs without brood. Within two or three weeks the rape will be in flower, and when the rape flow starts some of the colonies are certain to put up queen cells with the intent to swarm. Two steps beforehand, namely equalising the colonies and giving them plenty of room, are a great help in deferring swarm preparations and in making swarm control measures more effective.

Colonies need to be given room in advance of requirements. Give supers before the queen has filled the brood nest. Give supers to accommodate the bees, and the bees will later fill them with honey if there is honey to be got. Giving supers too late is a sure way of getting queen cells. If incoming nectar has to be stored in the brood chamber because there are no cells for storage elsewhere there will soon be no room for the queen to lay and queen cells will be started.

For some years now I have given colonies two or three supers, not just one, at the first operation. Immediately above the excluder I give a box of drawn comb, above that I give foundation or alternate drawn comb with foundation, with drawn comb on the flanks, and above that (if I give three, as I occasionally do) I give drawn comb. Although colonies at the rape will draw foundation very readily (and very well), and indeed I use the rape flow to get new brood combs built, they need comb, not foundation, in the supers from the start.

In very different circumstances, on my much less frequent visits to my Tyne Valley bees, I usually put two or three supers on each colony at the spring visit. I have no reason to believe that it would have been better to put them on one at a time, even if there were no practical difficulty in doing so, and I was interested to read, very recently, of American research which also reached this conclusion.[3]

Queen cells

To ensure beyond doubt that no occupied queen cells are present it really is necessary to shake or brush every comb in the brood chamber almost free of bees in order that none shall be missed. With practice, and with concentration on the job in hand and on nothing else, it needn't take long. Gregg suggests that ten minutes is ample for a colony on 22 combs with supers and that many would take less time than this.[4] I agree. It is an operation that requires speed but not rough handling, and the necessary quiet, smooth, but rapid handling without jarring comes only with practice. Speedily and carefully carried out the operation gives rise to no aggressive behaviour by the bees, indeed quite the reverse as a rule.

An operation such as this is vastly easier as a two man operation than on one's own, however experienced one is. But that is true of many beekeeping operations.

If and when queen cells are found, what then? I think there is no doubt that if a colony is found to have sealed queen cells some form of increase must be made. Even if the cells are less advanced than the sealing stage I think it is usually best to make increase, perhaps temporarily by taking a 3 or 4 comb nucleus. Learn to find queens, and learn thoroughly some well tried and reliable swarm control methods so that one or other can be put into operation without much hesitation. It is not knowing how and what to do that makes swarm control difficult and operations inappropriate. It is just not practical to have a book of instructions in one hand and the hive tool in the other.

Some beekeepers seem to direct their efforts to manipulations designed to frustrate the evident intention of a colony to build queen cells, or to operations which have that effect. I consider frustration methods unwise and unprofitable. In my view it is far more rewarding to pay careful attention to the strain of bees kept; to do nothing to prevent queen cells being built other than to give ample room for breeding and storage; and if a colony is nevertheless determined to build queen cells, to satisfy this urge in one way or another.

The variety of swarm control methods advocated, and the complexity of some, is bewildering. Take the trouble to learn thoroughly how to make an artificial swarm and a shook swarm, and some variations on that theme, and beekeeping becomes much simpler and more rewarding. Swarming, and swarm control, spell work for the beekeeper and a period of relative idleness and cessation of wax production for the bees. Artificial swarming and making nuclei take time. So do try to reduce the need for swarm control to a minimum, and above all keep bees that are reluctant to swarm.

Artificial swarming

An artificial swarm is made when queen cells with advanced larvae are found in a colony. The procedure requires the queen to be found, but for those beekeepers who cannot find queens, and who fail to find the queen, a modified procedure can be followed, as I describe later.

Proceed as follows: Remove excluder and supers, and set them aside on the upturned roof. Remove the brood chamber from its stand and set it aside. Put in its place a new floorboard and brood box containing empty combs, or say two combs and the rest foundation, leaving space in the middle for one comb. Find the queen in the moved colony and put the comb on which she is found, with the queen and the adhering bees, in the space left for it in the new box, between the two empty combs, first destroying any queen cells on this comb. Replace the queen excluder and supers, and the roof.

The moved colony is now queenless, but has queen cells. Select one queen cell containing an advanced larva and carefully destroy (without shaking) all other queen cells on that comb. Replace the comb in the hive. Shake all other combs free of bees in turn, destroy all queen cells on these combs, and replace them. Add a cover board and roof. Put this colony on a new stand.

The moved colony has all the brood, except one comb, has no queen, but has one advanced queen cell. It will retain all the young bees (except those taken with the queen) but will lose most of the foraging bees to the colony on the old stand, which has the queen, a little brood and attendant bees, and the bees in the supers. Three weeks later a new queen should be laying in the moved colony, and if increase is acceptable that is the end of the operation.

If increase is not desired, leave the young queen to lay for a while, then find and kill the old queen in the artificial swarm and unite the two colonies. Put the young queen and her brood box above a sheet of newspaper on top of the colony from which the queen has been removed, add a second sheet of newspaper and replace excluder and supers.

Other variations can follow the original operation, such as placing the moved colony to one side but close to the colony on the old stand, moving it ten days later to the other side, thus losing more flying bees to it, and uniting under the young queen when she is mated and laying.

Instead of being moved away, the box in which the queen cell is left can be put above the supers on a split board that provides an entrance to the rear or to one side, for the new queen to mate from there, and later united to the colony below or moved away for increase.

When the queen cannot be found at the start of the artificial swarming operation proceed as follows: Put with the two combs and foundation in the new box a comb that has very young grubs suitable for queen rearing, and no queen cells. Put the parent colony on top of the supers above a split board with a rear entrance after destroying all queen cells. Leave alone for a week.

If the queen is in the top box she will resume laying there and queen cells will be raised in the bottom box. If the queen is on the comb transferred to the new box the situation a week later will be as originally intended, except that new queen cells will have been raised in the top box as no queen cell was left there.

146

Diagram 9   Artificial Swarming

Move aside colony w qc & put new Ddc & f on old site. In centre of new D put c w q & add here bs from moved colony, flank w dc & g dc & f to fill. Replace XS (if any).

Move colony w qc to new site, & k qc, leaving one only .

S	
X	
New D	c w q & adh bs + dc & f to fill + flying bs

New D	k qc leaving one move to new site

OR, leave thus:

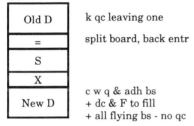

Old D	k qc leaving one
=	split board, back entr
S	
X	
New D	c w q & adh bs + dc & F to fill + all flying bs - no qc

OR, if q cannot be found, leave thus:

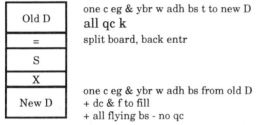

Old D	one c eg & ybr w adh bs t to new D all qc k
=	split board, back entr
S	
X	
New D	one c eg & ybr w adh bs from old D + dc & f to fill + all flying bs - no qc

Shook swarming

If increase can be accepted and the beekeeper can provide a site out of flight range of the colony with queen cells, even temporarily, a shook swarm can be made. A box suitable for transporting bees is needed. Use either a swarm box (a useful piece of equipment both for this job and for taking and transporting natural swarms) or a screened brood box (with or without combs).

First find the queen and shake her and the bees on her comb into the travelling box. Destroy all queen cells on this comb and replace it, clear of bees, in the hive. Next select an advanced queen cell on one of the other combs and destroy, without shaking, all other queen cells on this comb; then put the comb back in the hive. From the remaining combs shake bees into the travelling box until there are sufficient to make a good sized swarm. Put the cover on and fix it. Then destroy queen cells on the combs shaken, and on all other combs except the one with the selected queen cell. Rebuild the colony with excluder and supers.

Take the shaken swarm to a site beyond flight range of the colony from which it was taken, and hive and treat it as a natural swarm. Put an excluder between floorboard and brood box to ensure that it does not abscond, and remove this excluder a week or so later.

Diagram 10   **Shook swarming**

Find q. Shake q & bs from her c into new Ddc. Select 1 qc for retention, k other qc on this c but do not shake. Shake bs from 5 or 6 c into new D. k all qc except one selected. Replace XS. Take new D w shook sw to new site.

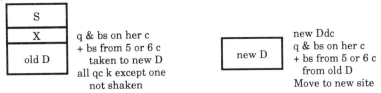

Taranov swarming

For those who cannot find the queen, or who cannot arrange a temporary site for the shaken swarm, an even simpler method of shook swarming is to use a Taranov board.[5]   I commonly use it in preference to other methods.

Set the board so that there is a gap of about 4 inches between the top edge of the board and the hive entrance, with both at a similar height from the ground, and lay a split sack or a sheet over the board to increase the area on which to shake bees and to ensure that the queen is not lost. Select a queen cell for retention and carefully brush, not shake, the bees off this comb onto the board. Destroy all other queen cells on the comb and replace it in the hive. Then proceed to shake all other combs in the brood chamber on to the sheet or sack. Destroy all queen cells except the one selected and replaced, and rebuild the hive with excluder and supers. Leave for half an hour or somewhat longer.

148

Most of the flying bees will return to the hive, crossing the 4 inch gap, but the queen and most of the young bees will not do so, and will cluster in the space below the board. Remove the sack or sheet when the cluster is fully formed, and carefully carry away the board, with its cluster of bees. The bees can be hived and treated just like a natural swarm, and, like a natural swarm, need not be taken out of the apiary.

I usually operate with a Taranov board slightly differently, as follows: Select and set aside, e.g. in a nucleus box, a comb covered with bees on which there is a good advanced queen cell. Make certain the queen is not on this comb. One way to be certain of this, if one cannot otherwise be sure, is to brush (not shake) the bees off the comb back into the hive and put the comb temporarily in a box above an excluder (on any colony) and wait a short while for young bees to come up and cover it, as they surely will.

Next move the brood chamber to one side, and in its place set a floorboard and an empty brood box. Then put the Taranov board in position and spread a cloth over it, or put the cloth down first. (The cloth is merely to increase the area on to which bees may safely be shaken, by covering the ground on each side of the board). Now shake all the bees, including the queen, on to the board. Destroy all queen cells on each comb as shaken and place the shaken combs in the new box on the old stand in the same order and facing the same way as before. Repeat with the combs in the second brood box if there are two, using the first brood box, now empty, as the upper brood box of the new hive for the reception of the shaken combs. Leave a space in the middle for the comb first taken (which will by now be covered with bees if it has been put over another colony), and put this comb, with its queen cell and bees, in this space. Add the excluder and supers, if any, that were on the colony, plus inner cover and roof. Then leave the bees to sort themselves out.

A Taranov swarm, just like a natural swarm, will be in good condition to draw foundation, if fed. I expect Richard Taylor would hive one on a shallow box of foundation and give it a super of round sections.[6]

Diagram 11   Artificial swarming with Taranov board

1. Remove & set aside SX. Select & set aside c w good qc but not w q. Move D aside.

Put new empty D in its place, & put c w good qc in centre. Then shake all bs, including q, on to Taranov bd. k qc as c shaken, & put shaken c in new D. Replace XS.

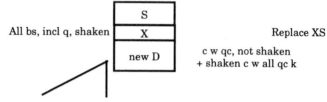

All bs, incl q, shaken

Replace XS

c w qc, not shaken
+ shaken c w all qc k

2. Half an hour or so later, carry Taranov bd to new site & hive sw.

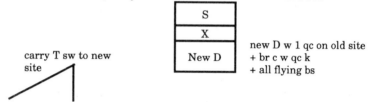

carry T sw to new site

new D w 1 qc on old site
+ br c w qc k
+ all flying bs

Other procedures.

A more drastic form of shook swarming is sometimes adopted, preferably at a slightly earlier stage of queen cell development, namely when there are queen cells with fat grubs but none yet sealed. Proceed as follows:

Remove excluder and supers (if any) and set these aside. Set the brood chamber (whether of one box or two) aside also, and in its place put a floorboard, and on this an excluder (not the one set aside) and then a broodbox with foundation in the centre and drawn comb on the flanks to fill. Shake all the bees from the brood chamber, (of one box or two) including the queen, into the new box. Replace excluder and supers (if any) or add both if there were none. Give the box (or boxes) of bee-less brood combs, after destroying all queen cells, to another colony above an excluder.

With a double broodbox colony the procedure can be varied so that the shook swarm retains one of the original brood boxes and combs, and this may sometimes be considered desirable, but it will then be necessary to check for queen cells a week or so later. Proceed as follows:

150

Put an empty brood box above the new box of foundation and drawn comb and from the upper brood box shake the bees from each comb into the new box, destroy queen cells, and put each comb cleared of bees and queen cells into the empty box, maintaining the original order. Then shake all the bees from the lower brood box into these two boxes, and replace the excluder, supers, cover and roof. Give the box of brood combs, cleared of bees and queen cells, (the former lower brood box) to another colony above an excluder.

Diagram 12   **Drastic shook swarming**

Remove & set aside SX. Set aside D. Put X on new floor in place & put new D dc & f on X. Shake all bs from old D, incl q, into (not in front of) new D. k qc on c as shaken & replace in old D. Put XS on new D. Put old D w shaken c & no bs on another colony.

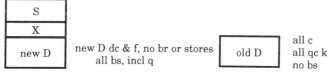

Whichever procedure is adopted, feed the shook swarm as for a natural swarm. Remove the excluder put on the floor below the new boxes from three to nine days after shaking, as may be convenient.

Making shook swarms, whether with or without a Taranov board, is not the difficult and hazardous operation that many seem to suppose, or that reading how to do it may suggest. Before shaking, smoke the colony rather more heavily than usual (more to encourage the bees to fill their honey sacs than to subdue them, so don't overdo it) and wait a minute or two. It is rare, in my experience, for shaken bees to be other than remarkably docile.

Artificial swarming with re-queening

Dr.A.L.Gregg suggested a method of artificial swarming with re-queening which I have used occasionally and found effective.[7] It requires the queen to be found, as a first step, and is, perhaps, not a method for beginners. Nor is it suitable for use in an out-apiary, as it requires a number of visits. Here it is.

If queen cells are found on a routine inspection proceed as follows: Find the queen, clip her wings if not already clipped, and put her in an old empty matchbox with three or four bees. Destroy all sealed queen cells and note the date on which the first virgin queen can be expected to emerge from those left. Rebuild the hive, but in doing so put the excluder on top. Above the excluder add a box containing three drawn combs and frames with foundation to fill, and into this release the queen. Add the cover board and roof.

If the colony has an excluder and super already then that excluder can be used, on top of the super, but a second excluder would allow the first to remain between brood chamber and super and continue to exclude drones from the super, and is desirable on that account.

Leave until a day or two before the first virgin queen is due to emerge, then take off and set aside the top box with the queen and the top excluder. Move the hive to one side, and put a floor board in its place. On this floor board put the top box with the queen, then the excluder, and above this add a super of foundation. Add a further box to contain the combs from the super on the moved hive and transfer these combs one by one, first shaking the bees off the combs in front of the new hive. Do likewise with any other supers on the moved hive. Add cover board and roof. Add a cover board and roof also to the moved hive, now reduced to its brood chamber of one or two boxes.

On the following evening, if queen cells are present in the upper box, (assuming there are two), of the moved hive, shake the bees off each comb from the lower box in front of the new hive, kill all queens cells on these combs, and put them in a box above an excluder on top of the supers of the new hive. Add a cover board or similar board above which provides one or two feed holes and an entrance above on each side. Cover the feed holes. On this board place the upper box of the moved hive, (or the one box if it is a single box colony), with bees and queen cells, and divide into two with a close fitting division board, making sure that there is a queen cell or cells in each half.

In due course a new queen will be laying, or probably two, in which case remove the combs from one half and treat it a nucleus. Allow the lot remaining to have the run of the whole box. Ten days or so later put a piece of queen excluder zinc over the feed holes in place of the solid covers, and nine or ten days later replace the cover board with a queen excluder. Ten days or so later, exchange the top box with the new queen for the bottom box with the old queen, and if all is well at the next inspection kill the old queen in the top box, shake each of the combs in the top box in front of the entrance, clear queen cells, if any, and replace the box of cleared combs on top.

Diagram 13   Artificial swarming with re-queening

1. When qc found, provide new D & rearrange colony thus:

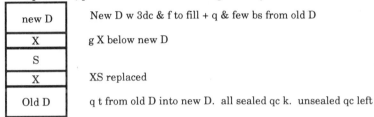

new D	New D w 3dc & f to fill + q & few bs from old D
X	g X below new D
S	
X	XS replaced
Old D	q t from old D into new D. all sealed qc k. unsealed qc left

2. A day or two before first vq due to emerge, exchange new D for old, shake bs off cs in S in front of new D, g Sf, replace S cleared of bs, and move old D aside for one day, thus:

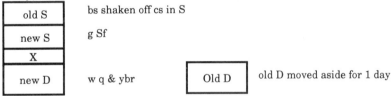

old S	bs shaken off cs in S
new S	g Sf
X	
new D	w q & ybr

Old D	old D moved aside for 1 day

3. The following evening, divide old D into two w qc in each half, & put D on top ,thus:

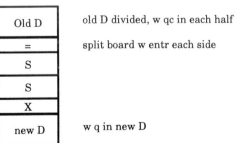

Old D	old D divided, w qc in each half
=	split board w entr each side
S	
S	
X	
new D	w q in new D

Using swarms

A swarm, whether a natural swarm or one made deliberately by the beekeeper, can not only be allowed or encouraged to build up into a new colony on its own, which it will usually do quickly and well with a little encouragement, but can also be used for a variety of purposes and in several different ways. You can use a swarm to strengthen a weak colony, or to re-queen a queenless colony, or a colony with laying workers. And you can, and should, use a strong swarm to clean up a dead hive with dead bees in the cells, or with wax moth. And for any of these purposes you can throw two swarms together to make one really strong one. Richard Taylor suggests hiving such a swarm in a shallow box and giving it two racks of round sections.

Natural swarms vary greatly in size. A really large one can weigh ten pounds and have 40,000 bees. Such a swarm, early in the year,. hived on its own and fed for a few days, will draw a full complement of frames fitted with foundation, or even only with starters, into comb in a matter of days, and will rapidly establish itself as a worthwhile colony for overwintering. In a good year it will get a super or two of honey. The old adage "A swarm of bees in May is worth a load of hay ..." indicates the value of a good early swarm. In contrast, a small swarm may have little more than a tenth of that number of bees, amd is best used to strengthen a weak colony or to re-queen a queenless one.

It pays to help a swarm to establish itself and build up. Feeding a swarm for a few days is common practice, and is generally desirable, but giving brood to a swarm when hiving it is less common, despite it being a great help to the swarm and ensuring that the swarm, when hived, does not abscond, as it sometimes will. Provide a box with a full complement of frames fitted with foundation in which to hive the swarm; remove the middle three of these frames and give them to a colony in exchange for three combs of brood. Choose a colony that can spare the loss. Shake the adhering bees back into the colony from which they are taken and put the three combs of brood into the vacant space in the box provided for the swarm. Then hive the swarm in the usual way.

If you can be sure that the queen is left behind, you can take the three combs with the adhering bees, thus helping the swarm still more, and possibly putting a desirable check on the colony from which the brood and bees were taken. If you do this, to ensure that the bees from the colony and the swarm unite peaceably, it is best to shake the bees from the three combs onto a sheet in front of the box provided for the swarm, and follow by shaking the swarm onto the sheet after putting the combs of brood, cleared of bees, into the vacant space in the box.

Using a swarm to strengthen a weak colony is best done by shaking most of the bees from the weak colony onto a sheet in front of its hive and then doing likewise with the swarm, so that both lots of bees enter the hive together and unite peaceably, and you can seek and kill one of the queens if you wish - a swarm queen can usually be found fairly easily as the swarm bees run into the hive - or you can leave the queens to fight it out, as they certainly will. Almost always (perhaps invariably) the swarm queen will prevail; she is slimmed down for swarming, her rival is not. So seek and destroy the swarm queen if you don't want her to head the colony.

An alternative way of adding a small swarm to a weak colony is to hive the swarm in a super, and add this box above a travelling screen to the weak colony, so that the two lots of bees have contact with each other only through the screen. Twenty-four (or not more than forty-eight) hours later quietly remove the screen and allow the two lots to unite, as they will then do peaceably. But remember that it is the swarm queen that is likely to survive, so you must first find and kill her if you don't want that.

Remember too, that a swarm queen is likely to be in her second (or her third) productive year, and a colony with her at its head will need to be re-queened with a young queen before the colony goes into winter. Remember too that late swarms may be second swarm (known as "casts") headed by virgin queens (and often with more than one). None the worse for that, but it should be known, and allowed for, so don't expect to see brood for a while from such swarms.

My bees in Essex are in a wooded area long known for its wild colonies of bees in the trees. It is from them that most of the swarms come that we see, and take. Until now it has been useful to have them, but I do wonder about their future with varroa around.

9 Oil Seed Rape

Lurid yellow shocks the eye.

Blunden & Turner, "Critical Countryside" (1985)

History and potential

Rapeseed and mustard seed have been cultivated for centuries for their oil, since before 1500 BC in India and since the 13th century in Europe. In the 17th century sufficient seed was produced in the UK to meet home demand and leave a surplus for export. Rapeseed oil was used mainly as a lamp oil, for which its slow burning properties make it particularly suitable. It has had a more recent use as an industrial lubricant.

The shortage of edible oils in the second world war led to the use of rapeseed oil instead of imported oils for the manufacture of margarine and to a four-fold increase in the acreage of rape for oilseed. The acreage has continued to increase, and rapeseed is currently in fifth position worldwide as a source of edible oil, after soya beans, sunflower seed, cotton seed, and ground nuts. In western Europe it is the most important oilseed crop. Canada, now the largest exporter of rapeseed, first grew rape for oilseed in 1942, to supply lubricants for marine engines.

Although grown in the UK for centuries, oilseed rape has only recently taken on a major role in the farming scene. Its role has been, and still is, as a break crop in cereal rotations. The choice of a break crop to maintain cereal yields is limited, especially if potatoes or sugar beet are not feasible, and rape both does well in the main areas of cereal production and is neither susceptible to nor a carrier of cereal diseases. Until recently it has not contributed much to farm income, but because of improved varieties, improved techniques, and a stable price for the product, it now does so, and it has the great advantages, in contrast to other break crops, that it is harvested by combine, as cereals are, and that it ripens before the cereal crops.

The area of oilseed rape in England and Wales increased from 16,000 ha in 1973 to 220,000 ha in 1983, and to 400,000 ha in 1993.[1] The area of rape now exceeds that of vegetables and potatoes, and ranks third to wheat and barley. Rape is now grown for seed in almost all the arable areas of the UK, including Scotland, but it is in eastern England, where two thirds of all the rape grown for oilseed is grown, that the most dramatic change has taken place.

Writing in 1945, F. N. Howes included Mustard and Charlock as major honey plants, but made no mention of oil seed rape either as a major honey plant or as one of the other plants visited by the honeybee.[2] In respect of charlock honey Dr. Howes pointed out that it granulated rapidly on exposure to light, even within three days, and often while still in the hive. This is, of course, equally true of the honey from oil seed rape.

Rape, as a farm crop for oilseed, can be either turnip rape, or swede rape, and either spring sown or August/September sown. In Canada, and in Sweden, oilseed rape is turnip rape, and spring sown. In the UK it is almost all swede rape, and mostly August/September sown. Plant breeding programmes, in Canada, in this country, and elsewhere in the EEC, have greatly changed the plant, and the composition of the oil, and this work continues. One aim is to improve the value of oilseed cake, the residue left after extracting the oil, as an animal feeding stuff.

In flower, rape is highly attractive to honeybees, and can provide nectar and pollen in abundance. It is not dependent upon insects for pollination and will set seed without such help, but some recent varieties seem less well adapted to wind pollination than others and we may yet see bees in demand as pollinators by growers. Farmers already like to have bees at the rape, particularly if it is autumn sown and spring flowering. The yield of seed is likely to be enhanced if the first flowers set, and the presence of honeybees as pollinators when the plants start to flower is therefore desirable and welcome.

Substantial areas of oilseed rape are now regularly to be found within flight range of many apiaries in eastern England and in the east Midlands, and not far away from many others, both in these areas and elsewhere. Both autumn sown rape, flowering in May, and spring sown rape, flowering in July, may be within reach, when adverse weather restricts autumn sowing or causes severe winter damage.

The potential for a substantial addition to the honey crop is evident. But the flowering period for autumn sown rape, from late April through May, is at a time of year when honeybee colonies are building up to, but will not have reached, full strength. Fields of rape in flower in May present beekeepers both with opportunities and problems.

Good colonies can get a lot of honey at the rape, but they need to be really strong if they are to do so.[3] More often the colonies build up on the rape, which they do with astonishing speed if they are literally in, or at the edge of, the field of rape, when they will work it in quite adverse weather.

In good weather bees will fly quite long distances to a field of rape in flower, but they will not fly even half a mile to it if the weather is poor. My apiaries in Essex are all within easy reach of rape every spring, but if the weather is poor, as it often is in early May, the difference between them directly relates to their distance from the rape. Colonies do best at, or in, the crop.

Fortunately, autumn sown rape can easily be identified by late October, and we consider what moves to make from that time onwards, find a site, seek the necessary permission, and set up hive stands. I set up hive stands in March, ready for moving colonies there. Indeed, I move to the rape almost all the colonies and overwintered nuclei that are not already within really close reach. Small lots, that have struggled to survive the winter, pick up wonderfully at the rape, if the bees have no distance to fly, and make good colonies for a later flow.

A well found double brood box Smith colony weighs 80 lbs. (36 kg) or more. I used to be able to lift and carry one, unaided, without undue strain, but I do so nowadays only when there is no reasonable alternative. Working alone, wintering most of my colonies on double brood chambers, and moving most of them to the rape, I had to devise a system within my capability, and I reduced them to single boxes for moving, usually in March.

By mid-March the winter cluster is usually established in the upper box, and at that time I lifted off and set aside the upper box (without inspection), then removed the lower box and set this aside (also without inspection) and replaced the upper box on the floorboard (or on a clean one). From each of the combs in the former lower box the bees were then shaken in front of the hive (the former upper box), and as soon as convenient the now single box colonies were moved to the new stands at the rape.

In mid March colonies can be moved to a new site within flying distance of the old site without significant numbers of bees flying back, as they have, by then, taken little more than cleansing flights. Colonies literally at the rape perform so much better than those some distance away that I commonly move half the colonies in an apiary within flight range of rape (as almost all mine are) but not actually adjacent to the crop, to a site at the edge of the rape field. By mid April such a move would lose most of the flying bees to the colonies remaining on the old stands. In mid March few bees fly back. By mid April, also, the brood nest would extend into both boxes, and the lower box could not conveniently be removed.

Occasionally, one finds brood in a lower box at the mid March operation; in which event the operation can be halted and the colony restored to its former double brood box state and left on its stand, or the box can be given to one of the other colonies left on the winter stand. The lower boxes removed usually contain some good store combs, and are used to accommodate and assist the development of overwintered nuclei, or in later queen rearing operation, or put back as supers on the single box colonies.

Problems

My out-apiaries in Essex have been within a variable distance, but always within easy reach, of rape grown for oilseed for the past twenty years. For rather longer I have taken colonies to the rape each year in numbers varying from eight to twenty. The experience teaches me to have three things very much in mind when the rape comes into flower, namely that the colonies may try to swarm, that they may suffer from spray damage, and that I shall have to extract the honey as soon as the flow ends.

Swarm preparations in some colonies are highly probable, and I need to have thought about what I shall do. Spray damage is possible, particularly when flowering is almost finished, and I need, at least, to ensure that the grower knows that my colonies are there and has my 'phone number. Granulation in the comb is certain, if I don't extract at once, and I need to set time aside for the job.

Swarming

Almost always some colonies at the rape will try to swarm. If and when they do it is no good giving them more supers, or foundation in the brood chamber, or even to take a 3 comb nuc with bees and brood and replace with foundation. With queen cells at an early stage at least replace the 3 comb nuc with drawn comb after killing cells. If a super of drawn comb, or of drawn comb with some foundation, is given at the same time the colony may be diverted from its swarming impulse. If it is not, more drastic measures are called for.

There is much to be said for taking a 3 or 4 comb nucleus from strong colonies at the rape as a routine practice. Replace the combs taken with drawn comb - possibly with newly drawn brood combs from an upper box, but not those with much honey. It is, of course, necessary first to find the queen, to ensure that she is not also taken. Many of the most desirable swarm control operations depend upon the queen first being found. It is a skill that has to be learned.

The ease or difficulty of finding queens varies greatly with the strain of bee. With some strains, particularly those that "run", queens can seldom be found. With others, queens take very little finding, because they quietly continue with their work.

If queen cells are more advanced, and close to being sealed, take the queen on the comb on which she is found, with the adhering bees, and two or three other combs of brood, to make a three or four comb nuc. Leave the colony to re-queen itself, or destroy queen cells and give a ripe queen cell from a selected colony.

Or take a 3 or 4 comb nuc with queen cells, but not the queen, replacing with drawn comb. Destroy all queen cells in the colony and give additional room, as much as possible in the form of drawn comb.

Or, more drastically, use Pellett's method of making increase,[4] as follows: Take the queen with her comb and adhering bees into a new brood box filled with empty combs. Remove the colony from its stand and set the box with the queen in its place, add an excluder and a second box (preferably a deep box) of empty combs, then the supers (if any) with a split board on top giving a back entrance, and finally the original brood box or boxes with queen cells on top. If desired, a ripe queen cell can be given in the top box 24 hours later (or at the time if protected). The supers can be given to another colony if the whole outfit is in danger of getting too tall for comfort.

Diagram 14   Pellett's method of making increase

Take c w q & adh bs into new D & g dc to fill. Move hive aside & set new D in place. Add X Ddc, replace S, add split bd w back entr & put old D on top .

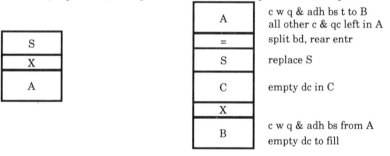

Alternatively, when queen cells are well advanced but the queen is still in the colony, make a shook swarm, or use a Taranov board and shake a swarm. Both operations are described in the chapter on Swarm Control.

Finally, you may find it useful to proceed much as we used to do in the apple orchards years ago, that is to move one of a pair of colonies away to a new site nearby, losing its flying bees to the neighbouring colony. One colony loses all its flying bees, and if queen cells are destroyed it is unlikely to build

162

more, but will make a good colony for a later flow. The other, remaining on its stand or moved to the middle of the stand the more readily to collect the flying bees from its neighbour in addition to its own, and perhaps with one or both of the supers from its neighbour added to it, will have a preponderance of flying bees and is likely to put honey in the supers rather than swarm, at least for a time.

Diagram 15   Moving away one of two adjacent colonies

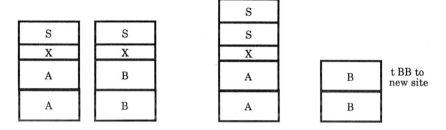

If I find one of a pair of colonies with queen cells and the other, perhaps somewhat less strong, with none, this is what I do. It is easy, quick, and effective, and does not reduce the honey crop.

Spray damage

It is not easy, and probably not possible, to avoid spray damage to honeybees at the rape completely. At least the problem is now far more widely appreciated than it was, and efforts to avoid it, by all concerned, are more usual. The extent of awareness and effort encourages the hope that we may reach a situation similar that in the fruit orchards, where spray damage to honeybee colonies is now quite uncommon, notwithstanding a continuing spraying programme.

Spraying oil seed rape to control insect pests, notably pollen beetle and seed weevil, is generally carried out only when the pest population warrants it, and not as a routine practice. The choice of spray material least toxic to bees, and the application of the spray early in the morning or late in the evening, or during cool weather, is also usually adopted, in an effort to minimise damage to honeybees. It is seldom necessary to spray while the crop is in flower.[5]

The best safeguard against spray damage that an individual beekeeper can take, in my view, is along three lines: first, to make yourself known to the grower and see that he has your 'phone number, and knows something of your problems; second, to learn as much as you can about why the crop may need to be sprayed, and when, and what damage may be caused to bee colonies; and third to consider what action you could take if you were warned of impending spraying.

Commonly, only one spray is applied, and that when flowering is almost over. The sense then lies in bringing the colonies home at once, but if they have supers, with honey in them, can you get the supers off, and the colonies home, in time? Sprayguard entrance closures, which permit returning bees to enter, but prevent bees from going out, are available, and effective, I am told. But if spraying is done in early morning, or late evening, as it can be, spray damage is unlikely to be serious. So do inquire.

Only once have I had colonies at the rape seriously damaged by spray, and that several years ago, from spray carelessly applied from an aircraft. Otherwise spray damage has been unnoticeable, or very slight. I think the experience of others is not very different, and that severe spray damage is, fortunately, not common.

Granulation

I keep more colonies than most as a spare time activity and in limited spare time, and can't really find the time to extract honey in June. But if my bees get rape honey in any quantity I take it off and extract it at once, and I do so not because I have customers queuing up for new season's honey but because I know it to be necessary.

Rape honey left on the hives for any length of time after there is no more to be had will granulate in the comb, sometimes very quickly, and what then? It is useless to try and extract honey from such combs, and it is folly to give a box of granulated honey to a colony as winter stores; the bees cannot use it, and may starve in the midst of plenty. It can be given to a needy colony in late spring, when the bees can freely gather water. They will then convert it into brood and bees, albeit somewhat wastefully. But only by melting the combs can the

164

honey be recovered, and this is a tiresome and tedious business, best avoided, if one can.

Beekeepers have told me, at Stoneleigh and elsewhere, that it doesn't happen to them; that they find rape honey granulating more slowly than I suggest. Sometimes, perhaps, but not in my experience, and I think it folly to act on that assumption. I don't even use Porter escapes to clear bees from the supers at the rape, because I find granulation well advanced in combs unattended by bees through a night or two. I take the honey off in the afternoon and start extracting that same evening.

By prompt and timely extracting the small scale beekeeper should be able to avoid granulation in the comb. Do so if you possibly can. If there are more supers to extract than can be completed in one session, try and arrange for those awaiting attention to be kept really warm, perhaps by the heat from an electric light bulb in an empty super below the pile, and don't delay longer than you have to.[6] Do the whole job of extracting in a really warm room. And don't delay in cleaning up; the traces of honey left on the extractor walls will granulate at an astonishing speed.

If you do find honey from the rape granulated in the comb and wish to recover the honey, there is no alternative but to melt it down. There are alternative methods. Cut the comb out of the frame, cut it up into pieces to melt on an uncapping tray, if you have one, heated sufficiently to melt the wax. Or pack the pieces into honey tins or buckets and heat these in a warming cabinet in which the temperature can be controlled until the honey is liquified. Set the control below the melting point of wax, say at 130 degrees F, and leave for 48 hours. Then use a spin drier to separate honey from wax, just as one does to extract heather honey by spinner. Allow the honey to cool a little before pouring it into a linen scrim bag, and spin modest quantities at a time, well within the capacity of the spinner. If one has no spinner, squeezing through a linen scrim bag between hinged boards would do the job, albeit slowly. Best buy an uncapping tray; it will serve two purposes.

Scraping the comb down to the midrib onto the uncapping tray or into a tin or bucket is an alternative to cutting the comb out of the frame, but is worthwhile only if the granulation is not far advanced. Even then it is a long and tedious business, and if the granulation is solid it is very difficult to avoid damage to the midrib to an extent that it is not worth retention for re-building into comb.

We get a lot of honey from the rape, and of late years we have just not been able to keep up with extracting the honey from the combs at the time. Of course we extract as much as we can, and extract through the night what we have taken off during the day, and put the supers back next day to be re-filled. But for several years now we have ended up with a stack of filled supers in which the honey is granulated solid. We don't worry. It becomes a late autumn job to get the honey out of these. We cut it out of the frames and melt it down.[7] We bought and use an outfit in which the chopped up comb is heated within a water jacket so that we can control the temperature at which it is melted, and keep it below the melting point of wax.[8] I don't like heating honey any more than I have to. We strain it warm and it ends up fit to bottle.

But we don't melt it all down. When we change floor boards in February we take a load of these supers with us and put one on the floorboard of any and every colony that is underweight when we lift it. So we save some for that purpose. Colonies will make good use of the granulated honey, if they need it, at this time of year, when they can get water to liquify it.

Willie Robson, with his thousand or more colony outfit in Northumberland, was a pioneer in dealing with the rape honey crop by melting it down. Briefly, the combs, newly built from starters, are set aside in a bee proof place until late summer. The combs are then cut from the frames, chopped up, and warmed to liquify honey and wax, for use in the normal way.[9] Today, heated trays specifically designed for melting granulated honey in the comb are readily available, and are both effective and efficient. And as to starters, we put the frames from which the comb has been cut for melting down back on the colonies in the following spring without foundation, or even starters. The remnant of comb left by the knife is quite sufficient to get the bees comb building on the rape much as one would wish. If they are going to be melted down again, as these are, some irregularity in the comb, and a degree of fragility, is no problem.

Of course, the rapid granulation of rape honey can be turned to advantage. Added to other honeys it will cause the lot to granulate quite quickly, and usually with a fine grain. Blended with other honeys it is a great help in presenting an attractive, top quality, granulated product.

A refractometer can be used to determine the percentage of water in honey, and is a useful tool to have and use when honey is to be stored. For safe storage the percentage of water should be 18% or less. I found pocket refractometers in use in Canada to determine the percentage of water in rape honey on the hives, so that the supers could be taken off when the percentage was in the low 20's% and stacked in a warming room with blown hot air, and extracted when the percentage had been reduced to 18%.[10] By this means granulation in the combs was avoided, and the extracted combs were returned for re-filling a bit sooner than they could otherwise have been.

Feeding

It is commonly overlooked, I think, that the removal of honey from the colonies when the rape is over may leave powerful colonies with very little stores, and a time of dearth to follow through a cold May and a June gap. In such circumstances the colonies will starve. A gallon of syrup to each colony on return from the rape is the answer, and is a wise precaution.

Likewise it is sensible to assist colonies to build up for the rape in areas where there are no early pollen sources such as pussy willow, (and perhaps even if there are), and I feed a pollen supplement, in the form of pollen patties, from mid February onwards. I deal with this in the chapter on Feeding Bees (Chapter 12).

Pollen traps

If pollen patties are to be fed, pollen has, of course, first to be collected, and this is done by means of a trap at the hive entrance which removes and collects part of the pollen loads of incoming bees. The operation appears to have no adverse effect on the colony, notwithstanding the removal from it, via the trap, of several pounds of pollen.[11]

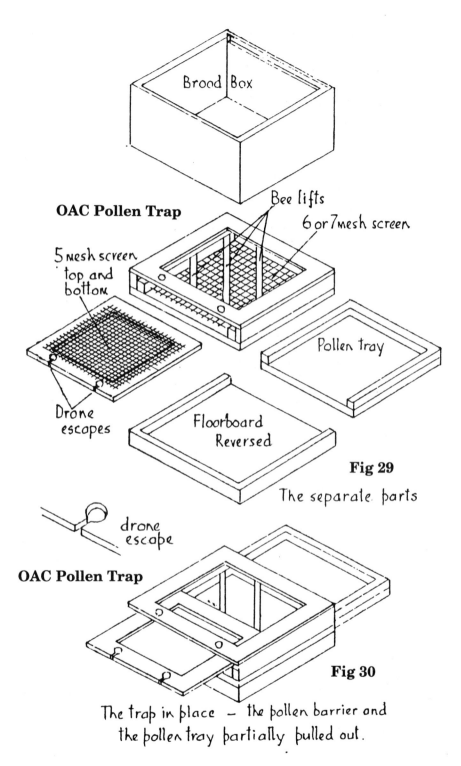

Brood Box

OAC Pollen Trap

Bee lifts

6 or 7 mesh screen

5 mesh screen top and bottom

Pollen tray

Drone escapes

Floorboard Reversed

Fig 29

The separate parts

drone escape

OAC Pollen Trap

Fig 30

The trap in place — the pollen barrier and the pollen tray partially pulled out.

The various designs of trap vary somewhat in their effectiveness. The OAC trap developed in Canada by the Ontario Agricultural College is widely used, and is perhaps the most effective and efficient. (See Fig 29 and Fig 30)

Nearly thirty years ago I made two OAC pollen traps, and trapped pollen, and fed pollen supplement for a year or two in Northumberland. Spring comes very late, if at all, in parts of that northern county. But it seemed to make insufficient difference to colony development - or to end results - to be worth continuing. My apiaries in Essex have rape in profusion within easy reach of the bees, and following Ted Hooper's advice and example I put these traps into use again, and made some more.

Each trap will collect enough pollen, if left on the hive through the season, starting with the rape, to make sufficient pollen supplement for five or six colonies, fed the following February onwards.

A pollen trap forces the returning foragers to pass through a barrier that will dislodge the pollen pellets from their legs. The OAC design of pollen trap overcomes the problems that can arise with these devices. It is easy to place in position, and can be in use for pollen collection or not in use, merely by inserting or removing a pollen barrier slide. The amount of trapped pollen is increased by using a double pollen screen rather than a single screen. Pollen is removed from the back of the hive without disturbing the colony, and can be left in the hive for two or three days without mould developing. It avoids congestion and crowding of bees at the entrance, and permits drones to leave the hive.

Installation and operation is simple. First lift the brood chamber off the floor board. Then reverse the floor board, putting the entrance to the back, and set the pollen trap on the reversed floor board so that the entrance to the pollen trap is in the original position of the floor board entrance. The pollen collecting tray slides on to the floor board from what is now the back, and slides out again in due course when there is pollen to be removed.

The tray is made of linen scrim, sackcloth, or similar material, fixed to a wooden frame, with supporting slats to hold the cloth 12 mm or rather more above the floor board. This frame should be a little less in width than the inside measurement of the floor board, and small staples driven into the sides of it and into the bottom of the sides, to project a trifle, will permit easy sliding. Air circulation around the tray will inhibit mould so that the tray need not be emptied more often than every two or three days.

At the bottom of the pollen trap unit a wire mesh screen (of 6 or 7 mesh/inch) prevents bees from gaining access to the pollen in the tray below. Three slats of wood placed diagonally and on edge are fastened to the upper side of this screen so that they just clear the pollen barrier screen above. These slats assist the bees to run up to the pollen barrier, and make for a more uniform distribution of pollen in the collecting tray.

The pollen barrier has two wire mesh screens (of 5 mesh/inch) spaced 6 to 8 mm apart by a plywood frame, and is constructed so as to slide in and out at will. Pollen collection can thus be started by sliding the barrier in, and discontinued by sliding it out, without any need to lift the colony off the trap. (See Fig 30)

Drone exits are provided by two holes of 25 mm (1 inch) diameter passing through both the top rim of the trap unit and the rim of the pollen barrier below, and two V-shaped notches connecting the holes to the front edge of the pollen barrier frame, so that drones can pass through and leave the hive. A piece of screen is tacked over the lower side of the hole and notch.

The outer surround of the pollen trap, which supports the hive above, is constructed of 19 or 22 mm material to match the floor board below and the brood box above, and the other parts are made of 6 or 8 mm material, largely plywood.

Pollen traps of generally similar design are available from appliance manufacturers. They appear not to have double pollen screens or the diagonal slat feature, both of which assist effectiveness.[12] Simple pollen traps for fixing to the front of a hive can also be bought, but they afford little or no weather protection to the pollen collected, and the tray has to be emptied daily (from the front) because of it small size.

Storing pollen

Fresh pollen can be frozen and stored in a deep freeze in paper or plastic bags, and will keep for several years. When

170

removed from the freezer it must be used immediately, or dried, or it will quickly deteriorate. Air dried pollen can be stored in an air tight container, such as a screw top glass jar, or a lever lid tin, but will gradually lose value and should be used within a year. A layer about half an inch deep can be dried in a honey warming cabinet or box in which the temperature can be controlled, at 120 degrees F for the first hour, reducing to 96 degrees F for 24 hours.

Alternatively, mix fresh pollen with half its weight of granulated sugar and store, covered with a layer of sugar half an inch deep, in an air tight container, such as a screw top glass jar. It should be used within a year.[13]

Pollen to eat

The composition of bee pollen is complex, but it is a rich source of aminoacids and of vitamins, particularly of vitamin B6, and of energy, and has healing powers. Many people eat it regularly and report significant benefits. I have seen it on sale in health food shops in Spain in attractive 500g and 1kg packs. and I expect it is similarly on sale elsewhere. Some easy ways to eat pollen are to sprinkle pollen grains on your morning breakfast cereal or yoghurt, or on honey spread on bread; or mix the grains with fruit juices or add grains to fruits in the blender or when using the blender to make salad dressing or cottage cheese.

"Lurid yellow"

For a few weeks in late spring the countryside in eastern England is bright with rape in flower. Too bright, some would say. Personally, I like it; but I know others who do not. It produces most valuable crops, both of vegetable oil and of honey. The public are beginning to connect rapeseed oil with oil seed rape. Perhaps we should now market rape flower honey by name, so that rape flower is also connected, in the public mind, with honey, its most attractive product. In France the rape plant is called colza and its honey is marketed as colza honey and is in good demand. To avoid the word "rape" the plant is now called canola in Canada and USA and rape honey is marketed as canola honey. Perhaps we should do the same.

172

10   Out-apiaries

If you must keep bees away from home,
keep them in a good producing district.

Frank Vernon, "Beekeeping" (1976)

Definition and desirable features

Even in the '30s my bees were in an out apiary. For more
than fifty years almost all my colonies - and for many years
literally all of them - have been in out-apiaries. I have never
had more than four colonies at home, and usually only two, and
a few nucs.

An out-apiary has been defined as an apiary site some
miles away from home.[1] It is that, of course, but it is more than
that. It is, in my vocabulary, a permanent apiary site away
from home. The qualification is important. It is, in my use of
the word, and, I think, in common parlance, a site on which bees
are kept throughout the year, and usually for a period of years;
on which colonies are wintered, and which may form the base
from which colonies are taken to crops for pollination, or to get
honey, as much as, and often more than, from the home base.

It is as well to be clear about the definition, because from
it follow most of requirements that determine whether the site
can fairly be considered a good one, or not. Temporary sites for
colonies taken to crops on a migratory basis are quite another
matter.[2]

From 1947 onwards I have had more than twenty out-
apiary sites in various parts of England. In all the early years I

173

had two, in Northumberland three, and in Essex four and now seven. They have always been selected with care, and I can recollect relinquishing only two other than for the reason that I left that part of the country. I have had several for ten or more years, and one for twenty-two years. So I regard the selection of an out-apiary site as deserving, and requiring, very careful consideration, and have clearly in mind what I think are the desirable features.

They can be listed very briefly, and then discussed more fully. They are: a good district, particularly for early pollen, and late pollen too, if possible; a good wintering site, particularly for wind shelter and winter sunshine; access for transport at all times; concealment, i.e. not readily seen by all and sundry; minimal nuisance to others; permission, and preferably welcome; distance from home.

They are not listed in order of importance. Except for the first, which could be dispensed with at a pinch, and the last, which could be stretched, they are all important and necessary.

The number of colonies in an out-apiary has some bearing on its selection, e.g. one might find a very good site that would accommodate only four colonies, but the number primarily depends upon intentions, rather than site. By intentions I mean the intended extent of migration to crops, which may, of course, be none. If so, the immediate locality, i.e. the area within flight range, is of first importance, and will determine how many colonies can profitably be kept on the site.

It is perhaps worth discussing the number of colonies in each out-apiary before considering site requirements. I tend to have fewer colonies in each out-apiary nowadays that in earlier years, and now have eight or twelve in each. In general, country districts are not prolific sources of nectar (at least in East Anglia) and are not as good as they were. But that makes migration to crops, e.g. to oil seed rape, the more important and the more likely.

So what transport facility does one have, and for how many colonies? How far is the site from home, or from another out-apiary? Will it fit the time available, one's speed of operation, and an acceptable round trip? Is the number of colonies likely to be sufficient to make the trip worthwhile?

Why establish an out-apiary at all? Presumably because it is either inconvenient to keep bees at home, or that only a small number can be kept at home and the desire is to have more, or that the home base is a poor one for bees and there are far better pastures not too far away. Or, as Wadey says in "The Bee Craftsman", in which there is a chapter on out-apiaries, "Some keep their bees in small town gardens under considerable difficulties, and in situations not the most productive, Others may be troubled by that greatest of beekeeping nuisances, the other beekeeper a few yards away. The remedy is to establish an out-apiary."

The answers to these questions will help in reaching decisions about location, and numbers, but leave us with the job of finding the site, and assessing its suitability. But I must add what ought to be self-evident, namely that keeping bees in an out-apiary is a job for an experienced beekeeper and not for a beginner, and that it is not practical without an adequate transport facility.

Good out-apiary sites are not easy to find. Don't seek permission until you have found one. You may not get permission (although you usually will) but you will be able, then, to say just why the site has been selected and to emphasise your consideration for others in making the selection. People are not likely to get stung because....and your operations will not interfere with nearby activities (farming or whatever) because....and you have given thought to this possibility or that. Always offer honey as a token rent - two 1lb. jars per colony is the going rate - and pay it in disastrous years as well as in good ones. Make a friend of the site owner, if you can, and keep him informed and interested. Few people know much about bees, and most are fascinated to learn.

It is always helpful, both to the bees and the beekeeper, if there are reliable sources of nectar and pollen within flight range of an out-apiary. Bees are always easier to handle and all the operations go more smoothly. With luck the site may prove good enough to make moving to crops elsewhere quite unnecessary. So look for such a site, and find one if you can.

Equally important, in my view, is an early spring source of pollen, perhaps from pussy willow, perhaps even from aconites and snowdrops. Generally, out-apiary sites in rural areas lack the early spring pollen sources of residential areas, which are very helpful to colony development.

Autumn pollen, at feeding time in August/September, is also helpful, perhaps from blackberry, and even later from ivy. Always bear in mind that the garden sources are not available out in the fields.

A good wintering site is of first importance. Look for wind protection, particularly from the east and north, and consider whether existing shelter can be supplemented. Make sure of winter sunshine - the colonies should be so sited as to get all there is. Look for good air drainage, i.e. not in a frost pocket, and not, positively not, under the drip of trees.

The site must permit vehicular access up to and alongside the colonies at all times, or at almost all times. You may be willing to walk a bit in the depths of winter, or in February when you may have a few pollen patties to carry, but don't even contemplate the possibility at other times of the year. You need to be able to pick up and load hives, and supers, and supply feed, without carrying any of these things more than a few yards. Anything else will make the whole operation a burden, and not a pleasure, particularly if yours is a one man operation.

It is not wise or sensible to choose a site where the colonies can readily be seen by passers by. It invites curiosity and vandalism. If the site is a good one in all other respects, consider whether careful siting of colonies on the site, and some screening, would help.

Consider particularly carefully possible nuisance to others, that is to all others going about their legitimate business. Is there a public footpath nearby? Is the farm track that you would use to get to and from the site also used by walkers? Are there houses nearby, and a likelihood of residents being stung? And consider, if the site is otherwise a good one, whether the possible nuisance could be overcome. Would screening, both to conceal and to throw flying bees up, meet the need?

The site, or rather the colonies on it, may need protection from livestock. Usually, a simple post and barbed wire fence will be sufficient, and provision for access is easily made. Discuss with the site owner.

176

The hives may also need winter protection against woodpeckers, which you have no means of knowing. It is best to assume that they do.

Put the colonies on hive stands, two on each. It is usually quite impracticable to keep grass cut in an out-apiary. It may be necessary to slash brambles, and cut grass around the hives, once a year, but hives on stands are not much inconvenienced by long grass beneath, benefit from the free movement of air around, and are at a convenient level for the beekeeper. Hive stands need to be reliable, so make a regular check. They should also be level.

Management

My colonies in out-apiaries are in groups of four, on two parallel stands about a yard apart, with each hive facing at a right angle to its neighbour. The arrangement minimises drifting, and is very convenient for the beekeeper's operations. It is an arrangement that I strongly recommend. Colony management in the out-apiary is altogether easier if hives and ancillary equipment are completely standardised and fully interchangeable. I consider it essential. In the home apiary one can have several different makes and types of hive without creating real problems; the bits and pieces to suit are near to hand. It is quite different in an out-apiary. Anything other than a single make of hive and full interchangeability of all the component parts will make the operation unnecessarily difficult and time consuming.

Never forget that you have to work in an out-apiary with what you have there and what you have brought with you. You can't often go back for something you have forgotten. It could be the matches. So you need adequate means of ensuring, without doubt, that you take with you what you will need. I keep tools, smoker, matches, etc. in a bee box,[3] and check that everything is there, and assess likely colony needs by studying my record book and making a "shopping list".

In an out-apiary you need a good, big, smoker, that won't go out, and plenty of fuel; a good pair of gloves; and wet weather gear in case it rains. You may have to continue you operations in the rain (when the bees are usually surprisingly docile!).

177

In an out-apiary also you need to adopt suitable methods of management, geared to weekly, or 9-day interval, visits, and not to more frequent operations. The convenience of methods that permit the division of colonies under one roof, and of having cover boards and split boards that can provide upper entrances, soon becomes apparent. So learn such methods and be ready to adopt them.

Remember also to take a travelling box and a nucleus box or two with you. It is often convenient to carry away a swarm or a nucleus, either to another out-apiary or to the home apiary.

Give supers early, two at a time. Make sure you have enough feeders of suitable size to enable you to give each colony in the out-apiary a two gallon feed. Feed in the autumn so that all you have to worry about is woodpeckers, and not possible starvation.

Finally, but most importantly, keep only the best bees you can - good tempered, not given to excessive swarming, and free from bad faults. Keep the colonies headed by young queens, deliberately selected, reared and mated, to achieve and maintain a really good strain well suited to the district.

Woodpeckers

Woodpeckers can do a great deal of damage in a very short time to beehives and to overwintering colonies. The green woodpecker is the culprit. These birds feed extensively on ants, and they attack beehives when anthills are frozen, but not only then; they acquire the habit, as the birds that take milk from bottles do. They are shy birds, that avoid people, are heard far more often than they are seen, and don't often move far from woodland. Out-apiaries in wooded areas are therefore particularly vulnerable.[4]

Typically, the birds attack the hives at the handholds, where the hive wall is less thick than elsewhere, and where two boxes meet. Left undisturbed, as they are likely to be in a rural out-apiary in winter, there will soon be a hole, or holes, an inch or two across, which may continue to be enlarged until it is bigger than one's fist. Colonies survive minor attacks, and even quite severe ones, but with large holes, and the destruction of combs which follows, the colony usually succumbs. In either event the damage to hive bodies is a costly nuisance, and not easy to repair.

Typically also, one can experience no trouble from woodpeckers for a number of years in an obviously vulnerable out-apiary, and then, one winter, find every hive in the apiary attacked. Perhaps the risk is small for some, but in a vulnerable situation it makes good sense to assume the likelihood of attack and take preventive measures.

The damage in my apiaries in Essex has all too often been severe. I have tried every preventive measure suggested to me. We had high hopes of fish netting, or garden netting, loosely draped around each hive, or each pair of hives. It can be bought off the roll, quite cheaply, in two metre or four metre widths, in whatever length you require. In the four metre width, a length of about three and a half metres is sufficient for each pair of hives, to enclose the pair and their hive stand down to the ground. Left "floppy", so that nowhere, except under the roofs, is it tight against the hives, the loose, floppy, netting should be an effective deterrent. Or so we thought. But wind caused the stuff to adhere to the hives, and we soon had holes bored through. Plastic sacks, cut open and pinned to the hives, had some deterrent effect, but not much. We now enclose each pair of hives with one inch mesh wire netting in late autumn. It works well, and costs much less than the damage to the hives and the colonies.

Frequency of visits

Of course, visits to out apiaries really need to be made at weekly, or at most 9 day, intervals from April to July, but it can't be done if the apiary is too far from home for frequent visits to be practical. For fourteen years, from 1971 to 1984 I had about 20 colonies near Hexham, in the Tyne valley in Northumberland, 250 miles from my home near Cambridge, and I had to devise a system of management suitable to a minimum number of visits.

It would normally make no sense to maintain an apiary so far from home, and it was the proximity to some of the best heather moors in England and the fact that the colonies there had always been managed for heather going that persuaded me to do so.

The apiary was eighteen miles from home when I lived in Northumberland. I had had bees there for eight years, and I left them there, temporarily as I thought, when I moved to Cambridge.

It is a near perfect wintering site, sheltered, getting all the winter sunshine there is, with an early pollen supply from aconites, snowdrops, crocus, helebores, pussy willow, wild cherry and apple, and so a good spring build up, and sycamore in late May. Aiming at the heather as the main crop, in August, it occurred to me that I didn't need to change my management practice very much to continue the operation on a minimum visit basis.

With colonies wintered on two brood boxes with really ample stores, I had never needed to visit before mid May. I had long operated on the Rauchfuss system of management, or very close to it, dividing all the colonies of or above average strength (and often all of them) in May and building the divisions up for the move to the moor in late July, and bringing the colonies back in early September. Five visits a year, in mid May, mid June, early and late July, and early September, could be adequate. I decided to try. In reality I had little option, as pressure of work in Cambridge didn't leave much time for beekeeping.

In fact it worked very well, although one had to accept some loss of operational efficiency. I lost a swarm or two, of course, but in the worst year no more than five, the crops of heather honey were satisfactory, and in some years there was a modest crop of flower honey too. I don't doubt that I should have had more heather honey if I had been able to give the colonies more attention, particularly in some years, and I had to accept that the operation did not pay; the cost of making a round trip journey of five hundred miles five times a year obviously outweighed the returns.

The aspect of the operation which bothered me most was my inability to influence or maintain the quality of the bees, which was initially very good, and the fact that I had to leave colonies to re-queen themselves with very little help (or interference) from me. Again, in fact, the colonies maintained themselves, and their quality, pretty well.

Except in one year, when I couldn't get there, (and all but one out of 26 survived), they had two gallons of feed on return

180

from the moor, and sometimes a gallon in June. They were always in good fettle. Two-gallon Rowse Miller feeders were left on as cover boards.

The continued trips north enabled me to keep in touch with friends there. Colin Weightman helped out from time to time, and Ernie Pope and I continued working together, which both made the operation possible and added to the enjoyment. It was hard work, particularly when moving to and returning from the heather, and age has inexorably caught up with us, so that I finally decided to bring the operation to an end.

When I lived in Devon I knew a beekeeper in Torbay who had four colonies in north Devon, sixty miles from home, which he visited twice a year, to put supers on, and to take them off. They were in double brood box Dadants and were always given two supers above an excluder. Twice I helped him take the honey off, and on both occasions they had a respectable crop of honey. He said they usually did.

I mention it only because I recently read of a French beekeeper who did much the same thing, also using double brood box Dadants, except that his colonies were ten times further away.

I don't recommend it, and I don't recommend my own foolishness, but evidently I am not the only madman in the beekeeping fraternity.

11   Heather honey

The production of heather honey is quite apart from the
production of other honeys. It requires a special study,
and only those who have a lot of experience
know the difficulties.

William Hamilton "The Art of Beekeeping" (1945)[1]

Among the many bee books that I have there is one,
published just after the war, with a substantial chapter on
heather going and heather honey in which the author rather
unobtrusively confesses that he has no personal experience of
heather going whatsoever. Some other authors, great names
among them, describe what heather going beekeepers do, and
should do, but don't seem to have much or even any experience
of heather going themselves. The genuine experience of other
authors is self evident, or expressly stated. Brother Adam, for
example, writes from an experience of taking colonies to the
heather in large numbers for many years.

Some informative articles by a number of experienced
heather going beekeepers have appeared in the bee press from
time to time, notably, and not surprisingly, in The Scottish
Beekeeper, and the Scottish Colleges have published useful
advisory leaflets. In the latter part of his tenure of office as CBI
for Northumberland, John Ashton did good work, with John
Theobalds, in analysing requirements, methods and results, and
was probably most successful in getting fully understood the
need for uninterrupted breeding at the right time and for really
powerful colonies at the moor.[2]

Retrospect

I took bees to the heather each year from 1950 to 1984, in small numbers in the early years, but for most of the time between 25 and 40 colonies, and in turn on the moors in Derbyshire, in the New Forest, on Dartmoor and Exmoor, and in Northumberland on the Durham border, above Rothbury and near Otterburn. I have had some very rewarding crops of heather honey and a few total failures.

My job took me to Derbyshire early in 1950 and I took a few colonies of bees with me. Most of the Derbyshire beekeepers I met took their hives to the moors each year, and I was easily persuaded to take mine. Neil Anderson's excellent scaled drawings of the Smith hive showed me precisely how to make some ekes,[3] and I had soon made four of these to fit inside two Smith brood boxes, and waxed starters of foundation to the bars. Beginners' luck, of course, but it is a fact that I had 50 lbs. of heather honey in the comb from one of the two colonies I took to the moor, and 51 lbs. from the other.

I vividly remember my wife and I cutting the precious comb, wrapping each piece in cellophane, and enclosing it in a section carton. We took this into Nottingham and I asked to see the honey buyer at Skinner & Rook's in the Market Square. He immediately told me that he had contracted to buy all the heather honey he required. I am not a good salesman, but I really did want his opinion on the product and pack that I had to offer. He at once said "that will sell like hot cakes" and asked how much of it I had. 101 pieces said I, and they really do weigh a pound apiece, which sections rarely do. They were on display in the shop that very day, I received more for them than I had expected (and kept one back to eat ourselves), and I was later told that the 100 pieces had been sold in three days.

I have never done quite as well since, and cut comb heather honey has become the usual rather than the novel form - but it does illustrate that a really good and attractive product will always sell, and explains why I have been "a heather man" ever since.

I was back in Kent two years later and took bees to the heather in the New Forest in 1953, '54 and '55, one year combining the trip with a caravan holiday with the family (when it rained most of the time!). I also had bees on Dartmoor in 1955 (I moved to Devon in that year) and on Dartmoor and Exmoor in the subsequent seven years.

The ling

Heather honey is, of course, the product of the ling (calluna vulgaris). The ericas, and particularly erica cinerea (bell heather), and in Cornwall erica vagans (Cornish heath), are also good and sometimes prolific sources of nectar, which have a flowering period rather earlier but often overlapping that of the ling. In Scotland there are large areas predominantly of bell heather from which good crops of honey are obtained. It can be, and is, extracted from the combs in the same way as other flower honeys, and is a beautiful reddish colour with a distinctive flavour. But it is not heather honey.

On some moors there is a good deal of cross leaved heath (erica tetralix) but bees seem rarely to work it. Some moors, predominantly of ling, have a significant admixture of wood sage (teucrium scorodonia) which flowers at much the same time and which bees work freely; and the ubiquitous willow herb is often within reach of the bees and in flower concurrently.

I have found the New Forest product a delightful honey with true ling honey characteristics in large part but undoubtedly far from pure in the sense of coming only from the ling, and the Exmoor product is very similar. The honey from the Derbyshire moors above Chatsworth was very good, and in the few years I had bees there a good sample of ling honey. On most of the vast areas of heather moorland in Northumberland there is nothing but ling heather, and the bees can therefore find nothing else to work, so that properly managed colonies can produce an unsurpassed heather honey, again in the sense of the absence of admixture of nectar from sources other than ling. There are large areas in Durham, in Cumbria, and in Yorkshire, and of course in Scotland, which I would expect to be equally good in this respect, but of these I have no personal experience.

Colonies of bees can starve at the heather, and often do. They need to be taken to the moors with an ample reserve of stores to guard against the very real possibility of a complete dearth of nectar if the ling does not yield. A common practice is to move colonies to the moor with two supers and a comb or two

185

of flower honey in the lower one to supplement the stores in the brood chamber. If the ling yields right from the start there will then be some combs with flower honey in them. The best beekeepers mark these combs and keep them for home consumption. Beekeepers who have a facility (control of temperature and humidity) to keep heather honey in the comb for a year, as I and a few others have, can put a comb of heather honey in the super, which probably has the added bonus of getting the bees quickly seeking nectar from the ling.

In terms of getting heather honey of good quality, stores of honey from oil seed rape in the hive are a menace. The least bit will cause rapid granulation of heather honey (which otherwise is very slow to granulate), and make storage in the comb impracticable. Actually the blend is very good, and much to most people's taste, but it needs to be made later, if at all, from extracted and pressed honeys. So, if the bees have been working oil seed rape earlier in the year, deliberate steps need to be taken to ensure that oil seed rape honey is not present in the hive when it goes to the moor.

The ling blooms over a long period, both because each plant is in bloom for a long time and because of variation in the time an individual plant comes into bloom. Aspect also plays a part, and experienced heather going beekeepers know the value of a site within reach of heather covered slopes that face the sun at different times of the day. Exceptionally, the ling may be in bloom on warm slopes before the end of July, even in Northumberland, and although the traditional date there for moving bees to the moor is still the week-end nearest to 7th August, I like to move them fully a week earlier.

Much of the heather moorland in northern England and in Scotland is managed for the benefit of grouse. A good grouse moor is a valuable asset that will produce a substantial income to the owner. Management for the grouse requires the heather to be burnt, in strips or blocks, on a rotational basis, at about eight year intervals, in order that there will be heather of all ages for the grouse to eat.

Ling heather is also an important source (and for part of the year the main source) of food for moorland sheep flocks. There is no inherent conflict between moorland management for grouse, sheep and bees, and there need be none.

186

Regeneration of heather after burning is largely from seed, and insect polination of the flower, predominantly by the honey bee, is essential to the production of seed. Young heather, from two to five or six years from seed, is quite evidently preferred by the honey bee and one can therefore infer that it yields the most nectar. Very old heather plants seem to yield very little.

I am in no doubt that a moor that is well managed for grouse is also well managed for bees. I could not be quite so certain about the sheep, as I think rotational burning over a longer period would suit them better (and might suit the bees just as well). But for choice, always take the bees to a well managed moor and not to a neglected one.

Moorland burning is not easy if it is also to be safe, so the interval does tend to slip. Burning can only be carried out during a short and restricted period of the year (outside which a special licence has to be sought and obtained) mainly to avoid damage or disturbance to nesting birds, and the number of days during that period when conditions are right, men can be got to help, and the job can be done, may be very few. Forestry Commission and privately owned woodland adjacent to heather moorland add greatly to the difficulty and the risk.

These woodlands often make excellent places in which to put bees for the heather. The colonies can be set on hard standings adjoining the forest roads so that little or no carrying is needed, and the shelter from the trees, at least during their early years, is an asset. The colonies are also then on private property, inside locked gates, and usually safe from theft. The Forestry Commission and private woodland owners both charge for the facility (some think they charge rather too much) and require common sense and consideration in selecting sites. I think this perfectly reasonable.

The Forestry Commission have recently taken to spraying heather to suppress or kill it in order to allow young trees to grow better with less competition. They give warning of intent, but the practice makes young moorland forests a much less attractive proposition to the beekeeper. The annual cash value of the heather honey crop in the UK must be very considerable, and could be very greatly more if more beekeepers took advantage of the vast areas of heather available, but the honey crop and beekeepers' interests seem to count for very little.

A degree of antipathy between the moorland gamekeeper, whose job is to produce grouse to shoot in a season that opens on August 12th each year, and the heather going beekeeper is understandable and only to be expected. It makes no sense to site colonies of bees near the shooting butts or near where the shooting party takes a lunch break, or even near a favourite picnic spot of urban visitors, and it makes good sense to come to sensible and acceptable arrangements with the keepers, and the moorland farmers, and sweeten the arrangement with a gift of honey.

Yields

Yields of honey from the heather can be quite astonishing. Ernie Pope, not at all given to exaggeration, tells of the 1949 season, when he and his mentor, Jack Mills, removed a full super of honey from each colony and replaced it with a super of foundation every week for five weeks and then had another to take off before bringing the colonies home; and that from the Newcastle Association's heather site on which there were close on 200 colonies, many of which did just as well. Colin Weightman records that in the same year he had 2,350 lbs. of heather honey from one group of 20 colonies on the moors above Blanchland, an average of 117 lbs. per colony. Brother Adam mentions a 95 lbs. average in 1933.[4]

But the heather usually behaves very differently, yielding copiously for short periods and yielding little or nothing for much of the time. Who can say why? It is certainly not wholly because of good or bad weather, or high or less high temperatures, or of the degree of wind.

The Norwegians (to whom the heather honey crop is of first importance) have provided some evidence that soil and air temperatures during the month or so before blooming influence the yield, and that generally the higher the temperatures then and during flowering the better the crop.[5] Maybe, but I and others have had good crops when the temperatures have not been very high and some less good crops when the weather has seemingly been better. I think night temperatures have something to do with it; an early night frost can bring a heather flow to a halt.

Colonies will sometimes fill and cap all the drawn comb in the supers but leave foundation quite untouched. This is usually the case when the flow of nectar is relatively poor. When the flow is copious the bees will draw foundation extraordinarily readily and fill and cap the combs as fast as you could wish. It would seem that the bees just can't help secreting wax at such times and will deposit small pieces of wax on much of the woodwork of hives and frames.

Beekeepers can do nothing about the weather or about the unpredictable nature of the yield from the ling. What they can do is to use a suitable strain of bee, take really strong colonies to the moor, and set them down sensibly and in good time on a good site. Brother Adam says that the two all important factors to success on the moor are the strain of bee and colonies of surpassing strength, and that is just what north country and Scottish beekeepers have been saying for years.

It seems likely that the nectar from the ling is not easy for the bees to work and process into honey. Certainly some strains of bee make a very poor job of it and are useless at the heather. Choose and stick to a strain with a good repute at the heather, particularly for its readiness to draw comb and cap well.

Nowadays we seek to produce the maximum proportion of the crop in the form of cut comb honey. The local strains of bee in the north of England and Scotland do the job very well, but I should add, in all fairness, that the ten colonies with bees of the Buckfast strain that I took to the Otterburn moor from Cambridge in 1975 did every bit as well. They produced a very worthwhile crop of beautiful cut comb.[6] I have since taken others, with starters in the frames, that have done equally well.

Management

A great deal of lifting and carrying can be avoided if the hives can be set down alongside the transport vehicle. But do not set them in a row or line. Set them in groups, or in a haphazard way, with entrances pointing in every direction, and minimise drifting. If you set them in a row you will discover the reason for this advice. The most populous colonies and the heaviest takes will be at one end of the row, and perhaps in the hives of another beekeeper who has added his two colonies to your line.

For the production of comb honey from wax foundation the hives need to be level from side to side, and a slight tilt forward will help to shed moisture. Attention to levelling really is necessary if you are after first quality combs.

I am sure that a source of drinking water is helpful to bees on the moors. A moist peat ditch is sometimes covered with drinking bees. Commonly the heavy dews of early autumn will serve, but if bees are to keep breeding on the moor they need water, and on some moors it is notably absent or hard to find.

I deal with the management of colonies for heather going in an earlier chapter, but it will bear repeating that colonies need to be really strong, with a brood box full of brood and stores, headed by a queen of the current or the previous year. I go with a single Smith brood box, excluder and two supers, some drawn comb in the lower super and foundation in the upper one, all unwired and extra thin for cut comb, and with a quantity of bees in the colony that the three boxes can barely contain. Nothing much less strong can be expected to get much of a crop.

Prepare the colonies the day before, close the entrances at dawn - 5 a.m. or thereabouts - load and travel and set down and level. Light the smoker, put on a veil, and prise each entrance block open just a little as you give a puff of smoke. When all the colonies have been released, retreat and enjoy your breakfast a little distance away. Finally return, again veiled and with smoker going, remove the entrance blocks completely (I put them on the travelling screen under the hive roof), make a final check that all is well, and pray for a good crop.

In good years some colonies will need a third super, and in poor ones they won't fill the two they set out with, but two full supers of heather honey is a very satisfactory crop (60 or 70 lbs). In my last years on the moors, because I lived so far away, my second visit was to remove the crop and take the colonies home. Do take the honey off the colonies at the moor. Comb honey on extra thin foundation is fragile - and valuable - and is really not well suited to a return journey on the hive to the wintering site.

I might add that heather going beekeepers also take to the moor quite small colonies and nuclei resulting from the year's queen rearing. They usually get their winter keep, and sometimes quite a small colony with a lot of brood and young bees, as these small colonies usually have, will get a super or part super of honey.

Finally, do remember that heather going beekeepers depend upon the goodwill of the moorland landowners, keepers, and farmers, and behave accordingly. In most heather districts such goodwill is traditional and readily forthcoming. It would be folly to lose it.

Extracting heather honey

Ling heather honey cannot be extracted from the comb by a centrifugal extractor, as other honeys can. It is thixotropic, i.e. more jelly-like than liquid in its normal state. If it is agitated, stirred, or squeezed, it temporarily becomes liquid and runs freely, and subsequently reverts to its thixotropic state. This characteristic is exploited by the different methods of extraction.

The traditional method of extraction, and still probably the most widely used, is by squeezing the honey from the comb in a press, in which the pressure is exerted via a pressure plate and a screw thread. Either the whole comb is cut from the frame and wrapped in linen scrim, through which the honey is pressed, or the comb is scraped from the mid-rib and the scrapings similarly wrapped and pressed. Pressing heather honey is a slow, tedious, tiresome and messy business. A good press is essential, and costly. But the product is good. If pure, or nearly so, it is very slow to granulate.

A few beekeepers use a hand held roller in which numerous needles are accurately set, to agitate the honey in the comb by plunging needles into the cells. The honey can then be extracted from the comb by a tangential extractor (a radial machine won't do it) in the normal way, and the combs saved for future use. The combs are first uncapped, of course, but fragments of wax in the honey are unavoidable, and have to be accepted, as they cannot easily be removed. Some large scale producers extract heather honey in this way, putting the uncapped combs through a power operated loosener and using swing-basket semi-tangential extractors.[7] (See Note 7).

A spin drier may also be used to extract heather honey, and many now use this method in preference to a press or an extractor. Both small and large machines are in use. The small scale producer, using a domestic machine, scrapes the honey and comb from the midrib, or cuts the comb out of the frame and chops it into pieces, into a linen scrim bag in a pail or similar container. The bag is tied at the mouth when sufficiently full and put into the spin drier. With a small domestic spinner it may be necessary to switch off after a few moments, when it begins to labour, and switch on again in a minute or so, and perhaps switch off and on again once more, to obviate labouring and damage to the drier.

The knack of using the drier for the purpose has to be learnt. Essentially it is to spin little and often. The method is very effective, and is quicker, easier, and less messy and tedious than pressing. It is best done in a warm room in which the honey in the comb has been standing for a short while. This has been my practice, but some users of the spin drier method of extracting heather honey warm the mush, scraped from the midrib or chopped up, for 24 hours at 110 F (43 C) before pouring it into the linen scrim bags for spinning. With care, this degree of heating probably does no harm to the honey and assists the extraction process.

Like other extractors, the spin drier needs to be elevated on a platform, so that the honey which pours out of the drier goes into a pail or other container. This is periodically emptied into a ripener to settle and allow air bubbles to rise, and for bottling.

But never forget that ling heather honey, unlike other honeys, cannot be strained. If particles of wax get into the honey they cannot be removed.[8] So press, or spin, heather honey in the comb through clean, strong, linen scrim straining cloth which will exclude and retain every little bit of wax. And never heat ling heather honey; its flavour, aroma and consistency is so easily ruined by heating.[9]

There is much to be said for producing heather honey in the comb, in the form of cut comb. Transparent plastic containers to hold a 6 or 8 oz. piece are available. It sells well, at a good price, and can be packed without mess (as the honey does not run) and with little or no waste. Treat it as a select product to a high standard. The pieces trimmed off, or only partly capped, can be pressed.

Commonly, all the pieces of comb trimmed off the chunks cut to size for the cut comb honey containers go into the pail and the bag and the spinner. Cut comb chunks are usually about four inches x two inches, and standard shallow frames (5½ inch or 6¼ inch deep) leave quite a bit for the press or the spinner. Alternatively, if one leaves a strip at the top the frames can be given back to the bees to clean out this strip and are then ready for re-use next year with a starter of comb in the frame. I have a number of supers 4½ inches deep made specially for cut comb, which leave little to press or spin.

Heather going - the relevant considerations

Since heather going is a somewhat specialised operation, and one which many beekeepers have yet to undertake, I summarise, and in some respects re-state, what I consider relevant, desirable, and necessary, as follows:

Ling heather honey - in comb or in jar - commands a high price. Demand exceeds supply. The nectar flow from ling is seldom a complete failure, although its timing is uncertain and unpredictable. It may come early or late in the flowering period.

A small patch of heather is of little use. Choose a site where a hundred acres (40 hectares) or more is within bee range. Locality, aspect, soil type, and elevation affect both quality and quantity of honey. At lower elevations more varied flora - willow herb, bell heather, wood sage - may be available, giving a heather blend rather than good quality ling heather honey. Higher elevations may have poorer weather with lower temperatures and strong winds. Slopes that afford more than one aspect will tend to prolong the flow. Nearby water - a burn, or moss - is desirable. A well managed moor where heather is burned regularly is to be preferred; young heather generally yields better than old.

Moors are usually windy, and a site that offers shelter should be selected. Prepare the site, and the stands, beforehand, if you can. Choose sites you can get to by car or truck to minimise carrying. Do not set hives in a row - the bees

193

will drift. Set them in pairs, or in a group facing different directions. Do not set them where shooting parties gather or casual visitors like to picnic.

To get a good crop of heather honey colonies need to be really strong in bees. When taken to the moor the colony should also have plenty of brood in all stages with a young and vigorous queen and an adequate reserve of stores. Colonies can starve on the moor if the flow is delayed. Don't blame the site or the weather for poor crops when the real cause is simply weak colonies. Bees wear themselves out quickly at the heather, and unless there are young bees emerging to replace loss, colonies will soon dwindle. 15 lbs. of stores in reserve is not too much.

Except where Dadant hives are used, best results are obtained from double brood box colonies reduced to one box by confining the queen to the lower box three or four weeks before moving to the heather. Brood boxes from which the queen is excluded will have the brood hatched out and at least partly filled with honey and pollen, and can be stored at home and returned to the colonies on return from the heather.

Use simple single walled hives. They are easily secured, pack well in the trailer or truck, and solid to handle. Travel with the entrance completely dark and with top ventilation; so block the entrance securely and provide a wire mesh screen on top. Securing hives for transport can be effected with hive staples, alone or with strapping; with strapping alone; with lock slides, alone or with strapping. Use a method which permits the addition of supers, etc., at the moor, and re-fastening for the return journey. Have some plasticine handy for emergency plugging of holes.

Don't travel at night. Mishaps are not easily dealt with in the dark. The bees may need to ripen nectar gathered that day. Travel as early as you can in the morning - really early. Make everything ready the day before and get up in time to shut the bees in at dawn. You can then load, travel, and unload in daylight, and enjoy breakfast on the moor. Release the bees (veil on and a puff of smoke) a little at first and fully after breakfast.

Moving bees by oneself is hard work and tiring. It is much easier with a helper (perhaps one's wife) and a pair of carrying irons or a hive carrier. Easier still if two beekeepers make the move together and help each other.

Some excitement of the bees is inevitable when a colony is moved. This tends to aggravate nosema disease, if present. It will pay to rid the apiary of nosema. Add Fumidil B to the syrup fed on return from the moor, and put all brood combs and boxes not in use through the acetic acid treatment.

Ling heather honey cannot be extracted from the comb like other honeys - it is thixotropic; more jelly-like than liquid in a normal state. If it is to be bottled it has to be pressed from the comb. Some form of press is needed. Wired foundation can be scraped to the midrib for pressing, preserving the foundation in the frame for future use. The scraped combs stored wet are very attractive to the bees the following season. Go to the moors with drawn comb in the supers if you can, and some extra thin foundation for the bees to make comb.

Pressing heather honey is a slow, tedious, tiresome and messy business. A good press is essential, and costly. There is much to be said for producing heather honey in the comb, particularly in the form of cut comb. In transparent plastic containers to hold a 6 or 8 oz. piece it sells well, at a good price, and can be packed without mess and with little or no waste. Treat it as a select product to a high standard. The pieces trimmed off, or only partly capped, can be pressed. A simple press such as that used to squeeze jelly bags (hinged arms) will suffice for a small quantity of freshly built comb, which presses easily. (See Fig 34)

For cut comb use shallow frames fitted with extra thin foundation (unwired). Manley type frames are ideal. The foundation can be partially drawn on an earlier flow with advantage. A starter of foundation about 1 inch in depth is all that many use.

Some prefer to use ekes. An eke is a box without top or bottom made to fit inside a super and provide a bee space all round between eke and super. Removable laths with a starter of foundation on the underside are lightly fixed in the top of the box. (See Fig 27). The bees build comb as in a shallow frame. No excluder is used, but a sheet of polythene is placed on the top bars of the brood combs, leaving about an inch all round for bee passage. It effectively excludes the queen from the eke. A number of ekes may be used, each separated by a polythene sheet. Two ekes in a standard deep box is usual and convenient.[10]

Comb is built in a vertical plane, so take care to set hives level from side to side, otherwise combs will not be built as you would wish. Take extra care when bringing the colonies home to handle hives very gently, and transport them carefully, to avoid the fragile comb breaking out of the frame or ekes. Preferably bring the cut comb supers home, cleared of bees, before bringing the colonies home.

Cut the comb into pieces to fit the plastic containers. A tool to cut pieces of suitable size can be bought, but a template and a sharp knife or cheese wire will serve. Include a suitable, small, "Heather honey" label, and state the weight, e.g. "minimum weight 6 oz."

Cut comb should be marketed directly it is packed. Retailers will commonly accept and pay for a carton of cut comb pieces in the plastic containers on the basis of their total net weight, and themselves weigh and charge for each piece. Do not cut more than you can sell immediately. If it is to be kept at all, (preferably not for long, unless temperature and humidity controlled storage is available), it is best kept uncut, i.e. in the eke or frame.

Colonies brought back from the moor benefit from being fed 10 lbs. or more of sugar as syrup, notwithstanding that the brood boxes are usually well filled with heather honey. Add Fumidil B to combat nosema. Queens usually stop laying at the heather and the feed gives an impetus for brood rearing to provide young wintering bees. Add the brood boxes that were left behind when the colonies went to the moor. Add these below the boxes that have come back from the moor if no feeding beyond 10 or 15 lbs. is intended, but above these boxes if the intention is to feed to fill.

**My car and trailer with some of Ernie Pope's Yarrow
hives, on a forest ride ready for unloading**

Hives set down for the heather on a forest ride

Ernie Pope with my bees, and some of his, on a heather moor near Otterburn in 1983

Colonies brought back from the heather
Note: Second brood boxes to be added, then fed for winter

12   Feeding bees

Some people seem to forget that the more they take away
the less there is left; as the Duchess said to Alice,
"The moral of that is, "The more there is of mine,
the less there is of yours"."

R.O.B.Manley, "Honey Farming"

Autumn feeding

Autumn feeding both ensures that the wintering colony does not starve, and greatly assists with its spring development. Colonies that develop rapidly in spring, and require the least assistance from the beekeeper, are those that overwinter with ample stores of honey and pollen and strong populations with a high proportion of so-called "winter bees". These are bees that were born in the autumn, and were well fed in the larval stage, but have had no nursing duties to do. They are old bees in terms of length of life but still young bees in terms of their physiological development, and fully capable of rearing the larvae of the first spring generation. Emerging from winter with adequate stores of honey and pollen such colonies are able to maintain relatively high rates of brood rearing despite external conditions unfavourable to an intake of much food.

The necessary conditions are most likely to be present in colonies that have been fed in the autumn while pollen is still available for collection. Given a young queen, such feeding will result in colonies entering winter with substantial reserves both of honey and pollen, a high proportion of "winter bees" and the ability to develop satisfactorily in spring.

The Manley type feeder has clear advantages, particularly in an out apiary. Provide one for each hive. You may have to make them yourself, so make them big enough to hold two gallons (or 10 litres). One feed will then suffice in almost any season and a second visit will not be necessary. Most of mine are of the David Rowse design, the Rowse Miller feeder, illustrated at Fig 8. This type permits the bees to enter and clean up the last drop of syrup, which is a feature that I like, but it does, or it may, make it difficult to re-fill without drowning bees, which is another reason for the feeder having a sufficient capacity to make re-filling unnecessary. Either make the feeders deep enough to hold two gallons or more, or see that the gap under the inner baffle is no more than 1/8 inch, when the bees can't get through to clean up and the feeder can be refilled without drowning bees.

Beehives are commonly set level from side to side and sloping down to the front about ½ inch, in which case it is best to arrange for the bees to have access to the feeder at the front. A Rowse Miller feeder 4 inches (100mm) deep for a Smith hive will then hold 10 litres (rather more than two gallons) of syrup, and will provide the equivalent of about 20 lbs. of stores at one filling.

The feeder can be left on for winter - not for feeding but as an inner cover - and will permit some top ventilation. It can be filled with vermiculite to provide insulation if you wish. These feeders - box feeders I call them - take up a lot of room in store, and I leave most of them on the hives over winter. But the materials and construction must take this use into account.

If you overwinter colonies in a single brood box and need to provide food quite early in the year you can give a block of candy or fondant on the top bars over the cluster and reverse the Manley feeder to cover and protect it. A month or two later you can reverse the feeder again and give a liquid feed. Working on double brood chambers I don't often resort to candy feeding (except in the form of a pollen supplement) but in late spring I sometimes give a liquid feed of a gallon or so. I think it best to avoid feeding syrup in spring until bees are flying freely and pollen is available for the bees to collect. In cold weather the bees won't take it from the feeder.

Except for feeding nucs, the 1 gallon plastic feeder, which has a piece of fine mesh copper gauze about two inches in diameter in the removable lid through which the bees take syrup, is the only other type of feeder worth bothering with. These have to be bought, and are quite expensive, but with care will last for many years. They have the advantage that you can see from the outside (through the translucent plastic) how much syrup has been taken.

Fill such feeders as full as possible, put on the lid and make sure it seats properly and is secure, invert the feeder over a bucket and give it a gentle squeeze or two or a shake to eject a little syrup, and it can then be put over the feed hole of a colony (don't invert it again of course) without syrup running out. Operate in the same way if you use lever lid tins with a few small holes in the lid as feeders.

Provided there are two feed holes well spaced apart in the inner cover, two of these 1 gallon plastic feeders can just be accommodated within the rim of the cover board, so you can give two gallons at one operation if you wish.[1] A good many of my inner covers (and all my escape boards) have two such holes. These are covered with a piece of glass throughout the summer and left uncovered in winter.

When Manley feeders are in use the inner cover sits on top and the roof covers all. With the plastic feeders I don't find any need to provide an empty box to enclose the feeder and support the roof. I don't have the empty boxes anyway. The roof sits on the plastic feeder and a brick is put on the roof directly over the feeder. Two plastic feeders sit nicely under the roof and one or two bricks makes all secure. But of course such feeders are not left on the hive very long. If they are left on empty, the bees will propolise the wire gauze, and this is not easy to remove subsequently so that the feeder may be used again. Perhaps boiling water is as good as anything for the purpose, but it takes time and is a nuisance. So avoid the necessity.[2]

Although these plastic bucket feeders serve their purpose very well, I should add that I have come to use them less and less, and scarcely at all for autumn feeding of colonies in out-apiaries. A single one gallon feed is only half the quantity I normally feed for winter, if I feed at all, so that if a single one gallon feeder is given, it has to be removed when empty and replaced with a second full one. In out-apiaries of eight or more colonies one soon learns that this is a tedious and time consuming business, even if one starts with a few spare full ones, as the empty feeders have to be re-filled before being replaced. Giving two such feeders at a time largely overcomes this objection but still leaves the necessity to make a further visit to remove the empty feeders before the wire gauze is propolised.

Manley or Rowse-Miller feeders are quick and easy to fill, and if need be to re-fill, and can be permanently on the hive or removed at any convenient time. Since these feeders hold two gallons or more, there is seldom a need to re-fill them, but if they are to be re-filled then do so before the bees have emptied them sufficiently to have access to the main compartment. The cover board over the feeder can then be slid, not lifted, to permit syrup to be poured in without releasing bees, and the operation takes very little time.

Feeders should be kept scrupulously clean. The plastic feeders are easy to clean (provided the gauze has not been propolised) and stack inside each other with the lids removed and thus take up minimum storage space. The Manley type feeders are less easy to clean, and stack like supers, i.e. one on top of another.

Feed in the evening if possible. The discovery of the sugar syrup usually leads to considerable excitement of the bees. Never feed honey; not only is there rarely any reason to do so but bees will waste seemingly endless time seeking an external source.

A strong colony of bees can take down syrup from a Manley type feeder with astonishing rapidity. I have known every one of a group of colonies to empty overnight, i.e. in no more than 14 hours, feeders containing more than a gallon of syrup, and to empty full two gallon feeders in 24 hours. Conversely, in really cold weather bees wont take the feed.
If there is pollen to be had, syrup feeding will stimulate a colony to get it. Many years ago Fred Richards advised me to use a 1 gallon lever lid tin with a few holes punched in the lid as a

202

feeder at spring fruit blossom time to stimulate the bees to collect pollen and do the job of pollination the beekeeper was being paid for. I went with him to buy the tins there and then, and most of them are still in usable condition. With care, beekeeping equipment lasts a very long time.

For carrying syrup to out-apiaries, and for making it in, I have found nothing better than 3 gallon (15 litre) home wine makers' fermenting bins. They are readily obtainable at a reasonable price, made of food quality plastic with tightly fitting lids. They are not too heavy to carry and pour from and travel without spillage. I melt 12 x 1 kg. bags (26 lbs) of sugar in 13 pints (7.3 litres; 16 lbs weight) hot water in each. Such syrup (62% w/w) is close to the highest concentration without risk of granulation in the feeder, and maximises its value as autumn feed.[3] For spring and summer feeding I melt 10 x 1 kg. bags (22 lbs) of sugar in 13 pints (16 lbs) hot water (59% w/w).

Until the advent of oil seed rape I did little or no feeding other than for winter in September. I was brought up on Samuel Simmins' adage: "Feed solid in September".[4] It is still very sound advice. In some years it was necessary to give colonies a feed in June/July to keep them going for the expected flow - and it still is. But I seldom, if ever, fed in the spring, and took no honey off until July. Oil seed rape has changed all that.

I have thought it best to deal separately (in Chapter 9) with the opportunities and problems that arise from oil seed rape, but here I repeat that the honey in the supers (and perhaps some from the brood chamber also) has of necessity to be taken off. So the bees get a gallon of syrup on return from the rape, as they do much later in the year on return from the heather.

Also because of oil seed rape, and the consequent need to encourage colonies to build up in early spring, I now feed pollen patties to the best colonies in late February, and I take round a few blocks of fondant as an insurance to others that appear light. The pollen patties are renewed in late March. By mid April, if there are colonies that need feeding, which is not often, they get a gallon of syrup.

Nowadays, I feed for winter a bit earlier than I used to do, and give colonies, other than those at the heather, a two gallon feed towards the end of August. It encourages queens to lay, and provide the necessary autumn bees. If the weather is good through September, as it often is, I may give a second feed of a gallon or two. Colonies on double brood chambers, as mine are, are not short of room to store feed, and I like to feed solid. They will then go through quite safely to the following May or June. It is always useful to have good food combs available to give to overwintered nucs, if need be.

Long experience has taught me that good colonies with good queens, well fed in autumn - fed solid, as Simmins would have said - not only winter well, but build up well and quickly in the spring. Indeed I don't know any better, or any other, way. We needed such colonies for the fruit blossom years ago, and we need them now for the rape.

Feeding pollen

I have found the feeding of pollen patties an interesting subject. They are messy to make and expensive to buy, and one can't be sure that they do any good, since one can't say how the colony would have progressed if the patties had not been fed. The bees take them readily, and they must, I think, be beneficial in removing from the colony the need to gather pollen in adverse weather, and thus working bees to death prematurely, particularly where there is not much pollen around for bees to collect.

Where there is ample pollen to be had around the apiary site, I think a gallon of syrup might be as good, or better. But who can say? Pollen, and pollen supplements, provide essential nutrients to enable bees to rear larvae, not a stimulus to the queen to lay eggs. Of course. a pollen patty may save a colony that is short of stores, and enable it to build up when it would otherwise dwindle, and perhaps die, but that it quite another matter.

The patties are made from natural pollen, collected in the previous year and stored, and expeller-type (fat free) soya-bean flour, in about equal quantities, mixed with sugar syrup to produce the consistency of thick porridge, or of a thinnish dough. There are several different formulations, usually with less pollen and added brewers yeast, eg., 20% natural pollen, 20% brewers yeast, 60% fat-free soya flour, and some with no natural pollen. No doubt they vary in effectiveness.[5] I deal with pollen traps and pollen collection and storage elsewhere (in Chapter 9). I have bought and used pollen substitutes (i.e. with no natural pollen) when I have been short of collected pollen and should do so again if need be.

My method of making pollen patties follows Johansson's basic recommendations.[6] I mix 1 part air-dried pollen with 4 parts water by weight, and when the pollen is thoroughly softened add 8 parts granulated sugar and stir until dissolved. Then I add 3 parts low-fat soya flour and knead into putty-like dough, which stands overnight. Next day the dough is made into flat cakes rather more than a pound in weight and about half an inch thick (about 500 g x 1 cm) rolled flat on to transparent plastic sheets or waxed paper. Wrapped in waxed paper or put into freezer bags, the cakes can be frozen and kept in the freezer until required.

The patty is laid directly over the cluster on the top bars, and pressed down between them a little, and covered with grease proof paper or a plastic sheet. I use a sheet of plastic large enough to cover most of the top bars. The rate of consumption can then readily be seen. Strong colonies will take two or three cakes, but I cease giving patties when natural pollen becomes available, and the weather permits the bees to collect it.

An alternative method of feeding collected pollen to bees is to feed it in empty brood combs without the addition of other food materials. Place an empty, or partially empty, brood comb on a flat surface so that one face of the comb faces upwards, and gently trickle or pour the pollen on to the empty cells in the comb. Lightly press or brush the pollen in to the cells with the fingers or palm of the hand, then turn the comb over and repeat the operation on the other side. Repeat the whole operation to provide more filled combs. Give one or two such combs to each colony to flank the broodnest, placed adjacent to the outside comb with brood, on one or both sides of the broodnest.

Bees take pollen given in this manner very readily and with evident benefit. It is a very effective method of feeding pollen, but it requires the removal and insertion of brood combs in each colony. I don't think it at all harmful to colonies to open them in February and remove a comb or two and insert others, provided one chooses a good day for the time of year, but it does take time. With colonies in out apiaries, as mine are, time is precious, and the provision of a pollen patty to each colony takes so little time that I don't feed pollen in the comb except to the few colonies that I keep at home for queen rearing.

Pollen supplements are quite widely used in USA and Canada, even in summer in desert areas where there is nectar to be had but an acute pollen shortage. There is evidence from New Zealand that colonies fed pollen supplement throughout the year perform better and produce significantly more honey than those not so fed. The colonies fed used an average of 5 kg of pollen supplement, and gave an increased return worth four times its cost.[7] Where there is a shortage of pollen in summer - which is unlikely in the UK - the continuous feeding of pollen supplement evidently might pay.

Feeding sugar and candy

It may be useful to know that bees take sugar - raw sugar, which keeps moist, not white granulated sugar, which goes hard - quite readily when they are active, as they are in spring. Put newspaper on the frame top bars, make a few slits in the paper, and pour the raw sugar on to it in a thin layer.

It has been suggested that 1 kg. bags of granulated sugar can usefully be given to the bees in mid winter, around mid-December, to ensure that single box colonies do not starve. The method is to take a 1 kg. bag of sugar and immerse it (unopened) in cold water for about ten seconds, then place up to four wet bags on the frame top bars where the clustering bees will gain access to the sugar by chewing through the paper of the bags.[8]

I should prefer to use candy or fondant myself. Both can readily be bought, and candy can be made. A cake of candy on the frame top bars, given around Christmas Day, and well wrapped, has been the traditional way of carrying single box colonies through the winter when otherwise they might starve. Provided the cake is renewed as necessary it is usually successful, and the second or third cake could, with advantage, be a pollen supplement patty. An empty super or a rim will be needed to enclose and protect the sugar bags or candy, whichever is used.

Colonies brought back from the heather are commonly heavy enough to come through the winter without feeding, but if they don't get a gallon feed on return (or perhaps even if they do) they should be checked in February by hefting, i.e. by lifting up the back a few inches to test the weight, and given a cake of candy or fondant if they are light. But one cake is unlikely to be enough when spring is late, so check again three or four weeks later and give another, and check again three weeks later and give a third if need be.

Baker's fondant is readily obtainable and not expensive. Get a 24 lb. pack and cut it into three or four pieces. Put each piece in a plastic freezer bag and seal it. It will keep in a freezer. Give one of these pieces to each colony that needs it, first making a good sized hole in the plastic bag, and put the bag of fondant directly on the frame top bars, hole side down, over the cluster. Cover with a sack or a piece of old carpet, enclose with an empty super or a rim, and replace the cover board and roof.

In Northumberland Ernie Pope and I used fondant in this way with colonies wintering on a single box on return from the moor, as I know many north country beekeepers do today, and I made a number of two inch rims to enclose the slabs of fondant; we had no empty supers. In so doing I was reminded of the very first hive I bought, a Langstroth, in 1927, which included, as a standard item, a two inch rim for candy feeding or for packing to conserve warmth.

Sugar fed bees

I have some reservations about feeding bees. Other than in the occasional disastrous season I have never fed sugar to an equivalent extent, weight for weight, to the honey harvested. Almost always it has been very much less; in fact, over a run of years, much less than half. Except when newly drawn out from foundation, and when culled for destruction and replacement, I don't extract honey from brood combs. Working on double brood chambers, as I do, there are usually several good food combs in the upper box, and sometimes upper boxes three parts full of honey, at autumn feeding time. These food combs are likely to have a good deal of pollen in them, and I have never thought it sensible to extract the honey from them, return the combs, and feed sugar so that the bees could replace it.

Those of us who keep our bees in out apiaries have progressively reduced the number of colonies in each apiary to match the local resources of nectar and pollen. I used to keep twenty colonies in each; I now keep eight. Manley writes of apiaries of 60 colonies,[9] and in my earlier years I knew of many with 30 and more. The total number of colonies may not be less, but they are now more widespread.

We ought, I think, to beware of feeding more sugar, weight for weight, than we take in honey. Dr Bailey has pointed out the clear correlation between high population densities of honeybee colonies in relation to available nectar and high incidences of contagious bee diseases. Beyond that number of colonies that is self sufficient in an area, that is when the number can be maintained only when sugar is fed additional to that required to replace any honey that is harvested, bees become underemployed, and diseases that are transmitted contagiously spread and multiply more than usual.[10]

Dr. Bailey concludes that the fundamental solution is not by suppressing these high incidences by therapy or by the selection of resistant strains of bee, but by appropriately decreasing the population density. In a natural desire to harvest as much honey as possible, beekeepers can easily fall into the trap of keeping more colonies than the district can support.

We may fall into this trap in areas where there is little or nothing to follow oil seed rape, unless we make a further migratory move.

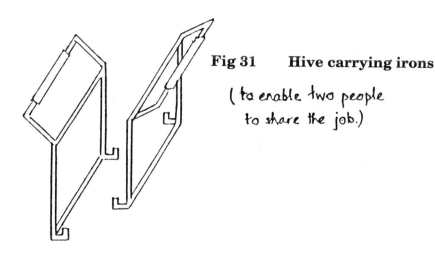

Fig 31 **Hive carrying irons**

(to enable two people
to share the job.)

Fig 32 **Bee Brush**

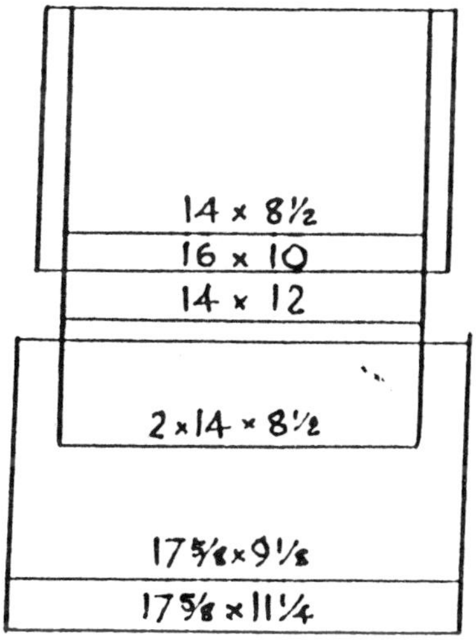

Fig 33 **Brood Frame Sizes**

B.S. : WBC, National,
            & Smith

14 × 8½

16 × 10   Commercial

14 × 12   British Deep

2 × 14 × 8½   Double BS

17⅝ × 9⅛   Langstroth

17⅝ × 11¼   Langstroth Jumbo & Dadant

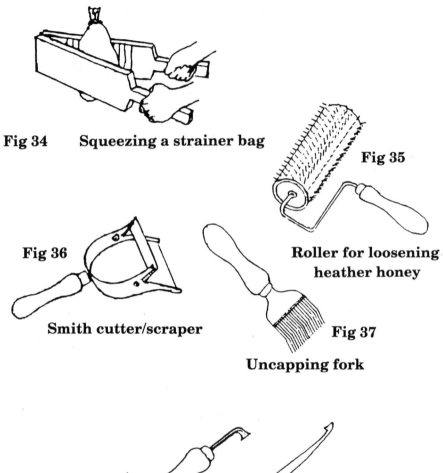

Fig 34   Squeezing a strainer bag

Fig 35

Roller for loosening
heather honey

Fig 36

Smith cutter/scraper

Fig 37

Uncapping fork

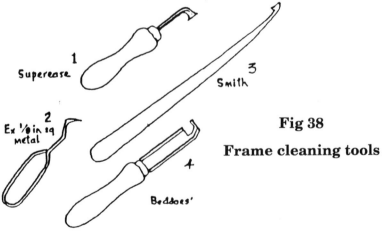

1
Supercase

2
Ex ⅛ in sq
metal

3
Smith

4
Beddoes'

Fig 38

Frame cleaning tools

An eke made up

An eke filled at the heather

211

Ernie Pope's beehouse and Yarrow hives outside
Note the lockslides

Ernie Pope's screen-board

13   Etceteras

It is not that these matters are afterthoughts.
Neither are they matters of less importance.
Like spices, they contain ingredients essential to success.

Ernest Bramah. "Kai Lung unrolls his mat."

Metric measures

Conversion to metric measures with complete accuracy can be
cumbersome and is usually unnecessary. The following are close
approximations:
1 oz. is approximately 28 g
1 lb. ..   ..   .. 450 g
2.2 lbs.   ..   .. 1 kilo
1¾ Imp pints (35 fl.oz.) = 1 litre
1 Imp gallon (8 pints) = 4.5 litres
1 fl.oz. is approximately 28 ml
1 inch = 25 mm
3 inches = 76 mm
degrees F - 32 x 5/9 - degrees C
1 kg to 1 litre is the same as 10 lbs. to 1 gallon, a suitable
strength of sugar syrup for spring feeding, when bees need
water.
15 kg to 10 litres water is a suitable strength of sugar syrup for
autumn feeding.
A 10 litre container holds 31 lbs. honey.

Disposal of the crop

For twelve years I bottled my honey for retail sale under my own label, kept a retail outlet supplied with my honey, and met a steadily increasing demand.

To maintain such an outlet requires the honey to be bottled, packed, and delivered on demand, when the retailer requires it, and an unfailing supply at any and all times of the year. Of course you can, and should, warn the retailer of a disastrous year, but the more nearly you can maintain a regular supply throughout the year, of honey of similar character, and quality, the better pleased he and his customers will be; and your own reputation will be enhanced.

You will probably be asked to supply both liquid and granulated honey and will find out the proportion of each that is required. Liquid honey must stay liquid, and granulated honey stay granulated, without becoming rock hard, or frosting too much.

It is not easy to maintain a similar character, colour and flavour in your honey from one year to another, which is undoubtedly what the customer likes. Stacking bulk containers of honey from different sources in known sequence and adopting the same blend each time when warming and bottling is the only way I know. Read what Manley has to say, in "Honey Farming".[1]

A warming box to take say 4 x 28 lb. tins or 30 lb. tubs and a 10 gallon bottling tank is the sort of outfit required. Put the selected bulk containers in the warming box one evening with the thermostat set at 100 to 120 degrees F, and they will be ready to pour into the bottling tank (in a warm room) next midday.

Blending and bottling

Left to crystallise on their own, some honeys will do so quickly and some very slowly, some with a coarse grain, and some with a fine grain but set very hard. What is needed is granulation with a fine grain but soft enough to be taken from the container with a spoon and spread on bread without difficulty. Only rarely, and unpredictably, will this be the result of doing nothing. What needs to be done is to stir into the liquid honey some finely grained granulated "soft textured" honey to spread the desired form of granulation throughout the sample - a process known as "seeding". Only thus can one ensure a product will be of the desired consistency.

On a small scale this can be done in the settling tank - the so-called "ripener" - at extracting time. When the honey has been extracted and strained and settled in the tank, skim the surface to remove debris or scum, then add some granulated honey selected as suitable for "seeding" and stir it in. Add between one and two pounds per one hundred pounds of liquid honey. Stir the mixture every day, gently and slowly, to avoid incorporating air, until it reaches a pourable consistency and a uniform degree of granulation. Then run it off into jars, or into bulk containers for later bottling.

Honey that has been allowed to granulate in bulk containers unassisted, and has an unsatisfactory granulation, should be warmed to liquify completely, poured into a settling tank and left to settle, and then seeded.

Honey granulates most rapidly at a temperature close to 57 degrees F (14 degrees C). For best results ensure that the honey to be seeded is at this temperature when the seed honey is added, and maintain this temperature, within a few degrees, during the mixing process; and keep the honey at 57 degrees F, or close to it, for at least five days after, by which time it will be set and will have a firm but spreadable consistency which it will keep if stored cool.

Granulated honey in bulk containers that has an acceptable granulation, such as honey "seeded" as above, should be melted before bottling only to a pourable consistency. Warm and melt such honey only to the extent that it can be stirred, and stir it while still warm. Warming should not go so far as to melt all the honey; leave a core of solid granulated honey to be incorporated by stirring.

When different honeys are to be blended, this is the time to do it. Pour the honeys to be blended into the settling tank, stir to mix the honeys thoroughly, allow to settle, and bottle while still pourable.

Provided care is taken to avoid incorporating air bubbles, and time allowed for air to rise and escape, a powered mixing propeller will ease the job of stirring. An electric warming ribbon around the settling tank, set at a low temperature, may also assist.

The use of a baffle tank to remove fragments of wax cappings, etc., from the honey drawn from the extractor can speed things up. Three or four baffles set so that the honey flows alternately over and under the baffles is what is needed. It can be home made; mine is. A double nylon stocking, suspended in a settling tank so that the stocking just reaches the bottom when extended by the honey within it makes an efficient strainer. The honey can then be run off from the settling tank into bulk containers or bottled for sale, and will be acceptably clean; indeed, as respects cleanliness, it will be of show standard.

When bottling honey, see that jars being filled are at room temperature or a little warmer, so that they are at a similar temperature to the honey. Handled and bottled in this manner the honey will retain a consistency suitable for spreading and freedom from frosting for a good long time.

Frosting in the jar is said to be indicative of pure and unadulterated honey, but it is not so regarded by the buyer and is not attractive or conducive to sale. It can largely be prevented by adopting a routine along these lines:

1. Have the jars at or close to the temperature of the honey, and warmer rather than colder. Do not pour warm honey into cold jars.
2. Do not fill the jars too quickly.
3. Run the honey down the side of the jar until nearly full.
4. Keep the jars at about 57 degrees F for five days, and at about 65 degrees F for a further five days.
5. Then store at about 54 degrees F until required.
6. At least, follow 1, 4 and 5 above.

Every jar of honey offered for sale under your label should be no disgrace to you at a reputable honey show. I used to send some to the major shows from time to time just to ensure that it was. The occasional prize card sent to the retailer always helped to boost sales, and had been genuinely earned by honey similar to that on sale.

216

Good honey, in good storage conditions, will keep a very long time. In the years just after the war I got to know a Dr. Oldfield, who ran a health clinic in north Kent. Honey was a basic food in his regime, and he was so convinced that honey improved with keeping that he kept it for a year before putting it on the table. I don't do that, but I always kept some from one year to the next, just in case the next year was a complete failure, and I still do.

It is not easy for a small beekeeper and a busy one to meet all these requirements, and when my retailer sold his business and retired, I stopped bottling for retail sale, except for friends, and except for cut comb heather honey, and sold all my honey in bulk, I have done so ever since. There is much to be said for it.

Cappings melter

I have had a large Pratley uncapping tray for many years, but have used it scarcely at all until recently. I have much preferred to uncap into a draining tank and melt the cappings later. With oil seed rape a principal source of honey it is so difficult wholly to avoid granulation in the comb before extracting that one turns to a heated uncapping tray almost in desperation. Large and efficient melters are now available that will accept whole combs cut from the frame, melt them down even if granulated solid, and separate honey from wax. Treated thus, combs granulated solid present no problem. But one is left with frames that have to be cleaned and re-fitted with foundation, which is both time-consuming and costly. I am also reluctant to heat honey more than I have to, so I bought and now use an outfit in which I can melt granulated honey in a water jacket at a controlled temperature.

Cut comb honey

Since the comb is cut to fit a standard container, the pieces are all cut to the same size, either by the use of a template or by a purpose made cutter. It is very desirable that all such pieces should have the same weight, or at least be closely similar in weight, and this can only be achieved if the combs are of nearly uniform thickness.

Manley frames, spaced by their 1 5/8 inch (41mm) side bars, are excellent for the purpose. The spacing is just about the maximum at which the bees will build comb without building wild comb between the foundation, and the frames should therefore be kept tight together. A wooden strip at one or both sides will both achieve this and make it easier, later, to remove the first frame. With grooved side bars and two piece bottom bars these frames are a standard item, readily obtainable. Fitted with thin foundation, unwired, the resulting combs are usually just what is required.

Don't fix the foundation in the frames much in advance of the supers being put into use, and don't give such supers to colonies much before the bees can be expected to draw the foundation into comb. It needs to be drawn into comb, and filled, rapidly and at once. Avoid rough handling of the supers at all times, particularly of supers in which comb is being built from starters. Cut comb is a choice, but fragile, product.

Provided the hive is quite level, and the nectar flow is good, starters of foundation in the same Manley frames may be adequate, and produce equally good, flat, uniform combs. Starters, about one inch deep, can be fixed by running molten wax along the top edge, or by fixing with the top bar wedge in the usual way.

Comb cut to fit the standard container from a standard Manley frame (5½ or 6 inches deep) leaves a substantial remnant in the frame. It can be used as the comb in a chunk honey pack, which usually sells well. The liquid honey around the pieces of comb needs to be light in colour and slow to granulate (i.e. not before sale). I have, and use, a number of supers that take 4 inch deep Manley frames, from which one can cut comb for a standard container without much comb remaining in the frame.

Except heather honey, which does not, honey runs from the cut faces of the comb, so that the cut pieces should be set on wire mesh to drain before being weighed and packed.

Cut comb should be sold and consumed as soon as possible after it is packed. It has a very short shelf life, so don't cut and pack more than the retailer can sell within, say, two weeks. Sale regulations require that the net weight is stated on the carton, so it has to be weighed. Minimum net weight is permissible, and may be prudent. The combs are best stored uncut in a warming box or cupboard in which the temperature can be controlled and maintained at about 90 degrees F. to retard granulation. There may be (and often is) a sale for whole combs, in the frame. Make it clearly understood that the frame is returnable, and see that it is returned.

Wild white clover, where it is still to be found, borage, and ling heather, are the sources that come to mind as being suitable for the production of cut comb honey. Except for a trial run on borage, my own production has been solely of cut comb heather honey. Honeys which granulate rapidly, such as those from brasicas (oil seed rape, mustard, etc.) and from raspberry, are not suitable for cut comb, unless granulated honey in the comb is acceptable.[2] I can see no reason why it should not be, but it would probably not be what the customer would expect and might give rise to complaint. If you know that comb honey offered for sale is likely to granulate quickly, it would be wise to make it known at the time of sale.

Round Sections

Ernie Pope and I tried a rack of Cobana round sections at the heather about thirty years ago.[3] We didn't get sufficiently good results to continue with them. Ten years ago I was persuaded to try again; Ross Rounds had become popular in the USA and Canada and could readily be bought, and Richard Taylor's books told one how to do it.[4] John Hunt had been pretty successful with them. At first I was not. In the area in which I live and keep bees there was no suitable and reliable nectar flow. One needs a really good nectar flow from a source of honey that does not granulate readily and has a pleasant, but not too strong, flavour.

The advent of borage as a farm seed crop entirely changed the picture. Honey from borage is light amber, bland, with no strong flavour but pleasant to taste, and is very slow to granulate. I have had tubs of unheated honey from borage that had not granulated by the following May. It is close to ideal for round sections if you can get the bees to work them quickly and well. The colonies need to be strong and to have nothing else to put honey in. And preferably no rape honey in the brood chamber to move upstairs.

When the borage comes into flower we move colonies there, and from them, and from colonies that are already close to borage, we take off supers, if they have any, and also take from the brood chambers what rape honey we can, and add boxes of round sections. Mine hold eight frames of four, and we usually give two to each colony to start with. We seldom get less than two filled, and we sometimes get four. But occasionally we get little or nothing.

In white clover districts, and there still are some, round sections could be well worthwhile. Perhaps also on the heather. Bees work them much more readily than they work square sections. But the equipment is costly, and although the frames, with care, will give many years service (we still have our Cobana racks in use) the rings and covers are given away with every sale and have to be renewed. The cost per round section of what you give away is as much or more than you may get for a pound of extracted honey, so do have that in mind when you put a price on a round section.

As Richard Taylor says, the idea of getting rid of all the extracting and bottling equipment other than a pocket knife and a table and a place to stack round section supers, is appealing. But I guess it would be impracticable except in a few favoured locations. It could be a very good way to keep bees for a beekeeper with no more than two colonies producing honey solely for home consumption and as gifts to friends; the rings and covers could be used again and again, as few would be given away. I might come to it myself in my very old age.

Winter work

Colonies of bees require little or no attention by the beekeeper through the winter; indeed they often require little or no attention for half the year - from October to April. In this respect beekeeping is quite unlike other forms of livestock keeping. But that is not to say that nothing need be done, other than dealing with honey, during these months. This is the time to do and necessary work in the workshop. I try and check, repair, and clean all equipment not in use, and make any equipment that I need. With a butane gas blowtorch I scorch all floor and cover boards, follower boards, and the inside of all brood boxes, as a disease precaution. Some deep boxes shrink, and should either be cut down for supers or to make feeders or have a fillet added to restore them to the correct depth. The maintenance of a correct bee space between boxes is essential to ease of management. Boxes that have been damaged by woodpeckers also need repair.

The winter is also the time to check stored combs. Culling and replacement of unsatisfactory combs is a necessary and important routine. Cut out such combs from the frames, and scrape and clean the frames (which can be scorched with the blowtorch) ready for fitting with new foundation in the spring. If you produce cut comb honey, scrape and clean the frames for re-use, and make up new frames if you need some.

Frame cleaning can be accomplished with a hive tool or a knife, but if the side bars are grooved, as mine are, buy and use one of the tools designed for the purpose. Four such tools are shown in Fig 38. I have all four. Find out what suits you best. It is a fiddly job, and should be made as easy as possible.

Fumidil B and Acetic Acid treatment

Nosema can be troublesome in colonies that are moved from crop to crop, and it is prudent to add Fumidil B to the autumn feed from time to time as a precautionary measure, and to fumigate all brood combs in store over winter with acetic acid.

To fumigate combs with acetic acid, the stack of brood boxes with combs needs to be on a floorboard with a solid entrance block that fits well, and covered with a well fitting cover board and a roof. Put a pad of absorbent material on the floor and pour about ¼ pint of acetic acid on it, and do likewise on each set of top bars. Cover, and leave for a week; then remove the pads and expose the combs to air (but beware of robber bees).

80% industrial acetic acid, which is what is needed, is nasty stuff, so wear rubber gloves. Bear in mind also that the fumes not only kill nosema spores but are very destructive of metal parts, such as hive runners, nails, and of course, metal ends. Remove what you can before treatment, and use plastic hive runners. Wax, honey, and pollen, are unharmed, and the combs, after airing, can safely be given back to the bees.

The fumes from acetic acid treatment of combs kill not only the organisms that cause Nosema and Amoeba but also the rearing stage of the EFB organism and the eggs and larvae of both kinds of wax moth. It makes good sense to sterilise all spare combs.[5]

Feeding Fumidil B

The instructions on the pack, to dissolve the powder in warm water and then add sugar, are best ignored, in my view; the powder does not readily dissolve. I find it best to mix the Fumidil B powder with dry sugar, add warm water (not above 50 degrees C) to make a syrup, and add this to syrup made in the usual way when the syrup is sufficiently cool.

Feeding it in sugar syrup is not the only way of getting the bees to take Fumidil B. Alternatively, give the stuff in weak solution via a 1-pint hand spray. Mix one third of a 0.5g vial (a 3-colony pack), or a dessertspoonful, with two tablespoonsful of sugar, and dissolve this in one pint of warm water (not above 50 degrees C or 120 degrees F). Add a further six pints of warm water. The seven pints so made will treat seven colonies. The method gets the Fumidil B inside the bees, and not in the winter stores, and a 3-colony pack treats seven colonies.[6]

The precise proportions are not critical, and for three colonies one could mix a teaspoonful of the stuff with a tablespoonful of sugar and dissolve this in half a pint of warm water; and add a further two and a half pints of warm water. The three pints so made will treat three colonies.

Start the treatment about three weeks before starting winter feeding. Spray the stuff over the tops of the frames (without moving them) and over the bees, and on the exposed parts of the four walls (inside) so that it runs down them. Do likewise for each box occupied by the colony. Repeat the operation after 6 or 7 days, and again after a further 6 or 7 days, making a fresh solution each time.

In emergency, colonies can be treated by the spray method during mild winter spells (when they can fly to evacuate the bowel) using much less than a pint per colony (say half a pint) and omitting to spray the walls. The full treatment can follow when the colonies break cluster in the spring.

In USA (and probably elsewhere) fumagillin is routinely fed mixed with dry powdered sugar. The proportions are not critical, and the contents of a 0.5g vial (a 3-colony pack) well mixed with 450g (one pound) of icing sugar and applied 50g at a time to three colonies, three times, at intervals of a week or ten days would approximate to current practice.

Terramycin

In western Canada and USA, and probably elsewhere too, Terramycin is routinely fed as a preventative measure for both European and American foulbrood. It is said to have a beneficial tonic effect upon healthy bees. I don't use it myself, but I know others who do. The practice is to give each hive one teaspoonful of Terramycin with one ounce of powdered sugar about mid November. Mix the dry ingredients thoroughly and then scatter the light orange coloured mixture over the top bars of the brood chamber. Alternatively, pour the dry mixture into a jar lid and slip it into the entrance at mid morning on a sunny day and remove it the next day. If some remains in the lid, give it back to the bees for one more day, first crushing it to a powder if it has become hardened.

In the USA it is illegal to feed Terramycin within four weeks of the beginning of a nectar flow; giving the stuff in November does the job without any fear of breaking this law. In the UK it is illegal to feed Terramycin to colonies known to have American Foul Brood, but it is not illegal to feed it to healthy colonies as a preventative measure.

In the Vancouver area of western Canada I discovered that some beekeepers use what they call extender patties between two brood chambers, using the recipe: 1 quart icing sugar: 1 pound Crisco: 1 teaspoon Terramycin. Mix the ingredients together to a consistency of soft butter. Form small patties (about 6 oz.) between sheets of wax paper, and put a patty on the top bars of the frames in the lower brood box towards the back of the hive, then replace the upper box. Renew the patties from time to time. It is said that by this method the medication will not and does not appear in the honey. Nevertheless the local beekeepers` association advises that the application of such medicated patties should stop at least three weeks before an anticipated nectar flow. I was told that some of the very large scale beekeepers in the USA routinely feed such patties each spring to their thousands of colonies.

Varroa

Varroa has not yet been found in my colonies, although it is known to be present in colonies not far away. Inevitably we shall find it in our colonies in due course. Meantime one learns all one can about methods to combat the varroa mite and checks for its presence.. Beekeepers in other European countries have learnt how to control varroa, and no doubt we shall also learn how best to live with it.

Killing bees

Very occasionally it is necessary to kill bees, and it is surprising how few people know how it may best be done. A safe and effective way (and the method routinely used to destroy swarms of African bees) is to spray the bees with a strong solution of household detergent (washing-up liquid) about half a cup of detergent to a gallon of water.

A hand held mist sprayer is effective, but a pressure sprayer with a lance, such us many of us have for garden use, will greatly increase one's reach, if it is an awkward swarm that has to be destroyed. For bees clustering with little or no comb, such as a swarm, the method can't be bettered. The spray will kill the bees quickly, and without fail, and except to the bees is quite harmless.

Making Mead

Talking with Brother Adam at Buckfast Abbey in 1956, and drinking his mead, I was persuaded to make mead, and I have done so nearly every year since then. Buckfast meads are very good. Brother Adam`s book "Beekeeping at Buckfast Abbey" has a section on mead, originally published in "Bee World" in 1953.[7] The article I wrote on Making Mead, in 1987, is at Appendix 2.

Bees in the garden

It is pleasant, and instructive, to see and hear bees at work in the garden, and I keep a colony or two at home as much for that reason as for any other. The absence of bees from other people's gardens is soon noticed. In aggregate, flowering plants in parks and gardens are often important sources, and sometimes the only sources, of nectar and pollen for colonies in urban and semi-urban areas. The contribution from plants in the beekeeper's own garden is necessarily small, and of limited value to the colonies kept there.[8] But in early spring the contribution can be of real value, by providing early pollen close at hand, and available for collection by the bees at every opportunity.

Plant early spring bulbs in quantity: winter aconites and snowdrops with crocus tomassinianus for February/March pollen. Plant puschkinia, scilla bifolia, and Dutch crocus to follow, and plant anemone blanda, scilla sibirica, chionodoxa, and lenten hellebores to continue the supply.

There are several dwarf willows that don't get too big, such as salix hastata wehrhanii and salix lanata, or the smaller weeping willow, salix caprea Kilmarnock, that will also provide early pollen. And most gardens will have room for at least one of the early flowering ornamental cherries, such as incisa, Okame, Kursar, and Hillieri Spire, or the slightly later Accolade, Umineko and Pandora, and the lovely Pink Shell, which are all garden size trees that don't get too big. So too is the dwarf crab apple, Malus sargenti, and maybe varieties such as John Downie and Everest.

Prunus x cistena, the purple leaf sand cherry, and prunus laurocerasus Otto Luyken, the low growing evergreen cherry laurel, are also garden size April flowering shrubs which bees love. So too is osmanthus delavayii, deliciously fragrant in flower, evergreen, slow growing (so choose a good sized plant to start with) and ideal for a sunny wall near an open window through which the fragrance and the hum of bees can be appreciated.

At this time, and into May, and later, bees in the garden are of service to the beekeeper as pollinators, if he or she grows soft fruit, or apples, or runner beans, or whatever.

In summer the globe thistles (echinops vars.), helenium, galliardia, and lavender always attract bees, and several annuals, notably limnanthes, phacelia, echium (vipers bugloss) and borage are sufficiently attractive to bees to be worth planting just to see the bees at work.

Late summer and early autumn flowering plants can also be of assistance to the bees in providing pollen at autumn feeding time. Caryopteris clandonensis, aster frikartii, michaelmas daisies, golden rod (solidago), ivy, and also the ivy-like Fatsia japonica and x Fatshedera lizei, are particularly useful and attractive to bees at this time.

My colonies on Dartmoor

Bibliography [1]

For Beginners

A Book of Bees, Sue Hubbell, Ballantine Book, New York, 1989
A Case of Hives, Len Heath (Ed), BBNO, 1985
Background to Beekeeping, Allan C. Waine, BBNO, 1975
Backyard Beekeeping, William Scott, Prism Press, 1977
Basic Beekeeping, Owen Meyer, Thorsons, 1978
Beekeeping, John Shida, Oxford, 1976
Beekeeping, H.R.C.Riches, Foyles, 1976
Bees at the bottom of the garden, A. Campion, Black, 1984
Discovering Beekeeping, Daphne More, Shire Pubs., 1977
*First Lessons in Beekeeping, C.P. Dadant, Dadant, 1983
*Honey Bees, A Guide to Management, Ron Brown, Crowood, 1988
*Keeping Bees, Peter Beckley, Pelham, 1977
Practical Beekeeping, Herbert Mace, Ward Lock, New Edn. 1977
*Teach Yourself Beekeeping, Frank Vernon, Hodder & Stoughton, 1986
*The Complete Guide to Beekeeping, Roger Morse, Hale, 1988
The Complete Handbook of Bee-Keeping, Herbert Mace, 1952, Revised Edn., Ward Lock, 1976
The Queen and I, Edward A.Weiss, Harper & Row, 1978

Practical Beekeeping

A Living from Bees, Frank C. Pellett, Orange Judd, 1945
A Year in the Beeyard, Roger A. Morse, Scribner, 1983
*Beekeeping at Buckfast Abbey, Brother Adam, BBPubl, 1974
*Beekeeping, Ron Brown, Batsford, 1985
*Beekeeping, Kenneth Clark, Penguin, 1951
Beekeeping in Britain, R.O.B. Manley, Faber & Faber, 1948
Bee-keeping, E.B. Wedmore, Foyles, 1946

Bee-keeping for Recreation and Profit, J.Harold Armitt, Ladbrook, 1952

Beekeeping Techniques, A.S.C. Deans, Oliver & Boyd, 1963

Beekeeping Tips and Topics, E.R.Jaycox, New Mexico, 1982

Bees & Beekeeping, Roger A.Morse, Cornell, 1975

*Guide to Bees & Honey, Ted Hooper, Blandford Press, 1976

*Honey by the ton, Oliver Field, Barn Owl 1984

*Honey Farming, R.O.B.Manley, Faber, 1945, facsimile reprint, NBB, 1985

*Honey Production in the British Isles, R.O.B.Manley, Faber, 1936

Practical Beekeeping, E. Tompkins et al, Garden Way, 1977

Practical Beekeeping in New Zealand, A.Matheson,GPP.1993

Principles of Practical Beekeeping, R. Couston, Kilmarnock, 1972

*Some Important Operations in Bee Management, T.S.K. and M.P.Johansson, IBRA, 1978

Swarm Control Survey, E.R. Bent, Gale & Polden, 1946

*The Art of Beekeeping, William Hamilton, Herald, 1945

*The How-To-Do-It Book of Beekeeping, Richard Taylor, Linden 1979

Queen Rearing

Contemporary Queen Rearing, H.H. Laidlaw, Dadant, 1975

Queen Breeding for Amateurs, C.P. Abbott, Bee Craft, 1947

Queen Rearing, Laidlaw & Eckert, California, 1962

Queen Rearing, Ruttner, IBRA, 1985

Queen Rearing in England, F.W.L. Sladen, 1913, facsimile reprint, NBB, 1982.

*Queen Rearing Simplified, Vince Cook, BBPubs, 1986.

*Rearing Queen Honey Bees, Roger A.Morse, Wicwas, 1979

Comb Honey

*Comb Honey Production, Roger A.Morse, Wicwas, 1978

Honey in the Comb, Carl E. Killion, Killion, 1951

*Honey in the Comb, Eugene E. Killion, Dadant, 1981

*The New Comb Honey Book, Richard Taylor, Linden, 1981

Bees, Honey and Plants

Anatomy and Dissection of the Honeybee, H.A. Dade, IBRA, 1962
Breeding Super Bees, Steve Taber, A.I.Root Co., 1987
*Breeding the Honeybee, Brother Adam, NBB, 1987
Breeding Techniques and Selection for Breeding the Honeybee, F. Ruttner, BIBBA, 1988
*Honeybee Biology, John B. Free (Ed), Central Assn.BK, 1984
Honey Bee Pathology, L. Bailey, London, 1981
*Honey Bee Brood Diseases, H. Hansen, Wicwas, 1982
Honey for Health, Cecil Tonsley, B.B.Pubs., 1969 and 1980
*In Search of the Best Strains of Bees, Brother Adam, NBB, 1983
Insect Pollination of Crops, John B. Free, London, 1970
Oilseed Rape & Bees, Allan Calder, NBB, 1986.
*Plants and Beekeeping, F.N.Howes, Faber, 1945, R.Edn. 1979
Pollen Identification for Beekeepers, Rex Sawyer, Cardiff, 1981
The Anatomy of the Honeybee, R.E. Snodgrass, NewYork, 1956
The Bee Community, F.H.Metcalf, Bee Craft, 1948
*The Behaviour and Social Life of Honeybees, C. Ribbands, London, 1953
The Honeybee, C.G. Butler, Oxford, 1949
*The Honeybees of the British Isles, B.A.Cooper, BIBBA, 1986
*The Pollen Loads of the Honeybee, D. Hodges, IBRA, 1952
The Social Organization of Honeybees, J.B. Free, Arnold, 1977

Special Interest

Beeswax, Ron Brown, BBNO, 1981
Beeswax, Coggshall & Morse, IBRA, 1984
Honey Marketing, H. Riches, BBNO, 1989
Lets Build a Bee Hive, Wilbert R. Miller, Miller, 1976
Making Mead, Roger A. Morse, Scribner, 1983
The Miracle of Propolis, Mitja Vosnjak, Thorsons, 1978
The Observation Hive, Karl Showler, BBNO, 1978

General Interest

A Book of Honey, Eva Crane, Oxford, 1980
A Nomad amongst the Bees, J, Johnston, NBB, 1995

*Beekeeping and the Law - Swarms and Neighbours, Frimston and Smith, BBNO, 1993

Beemasters of the past, V. Dodd, NBB, 1983

For the Love of Bees, Lesley Bill, David & Charles, 1989

*Fifty Years Among the Bees, Dr. C.C. Miller, 1915, (facsimile reprint, Molly Yes, 1980)

Honey, Isha Mellor, W.H. Allen, 1980

Honey Days, Oliver Field, N.B.B., 1980

New Beekeeping in a Long Deep Hive, Robin Dartington, BBNO, 1985.

Tales of a Border Beekeeper, Colin Weightman, Ltd.Edn.1991

The Art of Beekeeping, 2 vols, O & H Aebi, Unity C.A., 1979

The Bee Craftsman, H.J. Wadey, Bee Craft, 1943

The Behaviour of Bees - and of Bee-keepers, H.J. Wadey, Bee Craft, 1948

The Border Bees, Colin Weightman, Williams, 1961

The Joys of Beekeeping, Richard Taylor, St. Martin's, 1974

The Philosophy and Practice of Beekeeping, A.L. Gregg, Bee Craft, 1945

Encyclopaedic

ABC and XYZ of Bee Culture, A.I. Root (Ed), Root, 1980

*A Manual of Beekeeping, E.B. Wedmore, Arnold, 1946,
   facsimile reprint 1975, reissued BBNO, 1979

*Encyclopedia of Beekeeping, Hooper & Morse, Blandford Press, 1985

Honey, a comprehensive survey, Eva Crane (Ed), Heinemann, 1975

The Hive and the Honey Bee, Dadant (Ed), Dadant, 1976

World Perspectives in Apiculture, Eva Crane, IBRA, 1986

Postscript

It all began with Brother Adam. I spent several weeks in Devon, intermittently, in 1980. I still have a house there. I had opened a major public Inquiry the previous November, heard evidence for eighty days, and closed the Inquiry in May. At my house in Devon I was able to give uninterrupted attention to the massive report I had to write. When it was completed it ran to 254 A4 pages and sixteen appendices. I took the opportunity of calling on Brother Adam at Buckfast, as I usually did when I was in Devon. He knew what I was busy doing. He urged me to write a book on bees and beekeeping as soon as I could find time. He knew that I had declined an invitation to write such a book twenty years earlier, but said that I should do so now, pointing out that I had a quite unusual experience of working with bees; continuously for nearly forty years in substantial numbers and always in out-apiaries, and in diverse parts of England, through Kent, Devon, Northumberland, Essex and elsewhere. I had, at that time, more than twenty colonies on a permanent site more than two hundred and fifty miles from home, as he knew. For the benefit of other beekeepers, he said, I ought to write a book; about bees, about colony management, and about what I had learned from such a long and varied experience as a spare time beekeeper. I said I would try.

I spent much of 1981 in the USA and western Canada, so it didn't get started until 1982, when I was recuperating after a short spell in hospital. I took the opportunity to sketch out the plan of the book, outline chapters, and see what material I already had. It got as far as a draft in outline. I continued to add notes as they came into mind, but it was not until late 1984 that I could find time to concentrate on the book.. My beekeeping friends, and my children, had long urged me to do that, and I got down to it as soon as I could. By then I had fifty colonies in Essex that I worked without help.

I had an unusually long session with Brother Adam in 1986, when we discussed a multitude of topics, including my book. He had seen two Chapters in draft (Heather Honey and Out-Apiaries) and the others in outline. A bit later I sent the book in draft to three of my friends, who liked it, and made some suggestions, and it was ready for publication, or very nearly so, by the end of the following year, 1987. But then fate intervened; my (Canadian) brother died, my wife of fifty years fell ill, needed care, and later died, and I had an accident and suffered a fractured skull, from which it took some time to recover. The book, on A4 paper and on disk, sat on the shelves awaiting attention.

I went to see Brother Adam in County Durham in 1994, when he was staying with an old friend and recovering from illness, and one of the first things he said to me was "Where has that book of yours got to? Why is it not yet published?" He was then a few weeks short of his ninety-sixth birthday. He urged me to see that it was published without further delay; an entreaty echoed by others. So, at last, here it is, dusted down and reconsidered for publication. But, alas, not in time for him to see and read,

He was a kindly soul, quiet and unassuming, but he set high standards and was a severe critic, so I can but trust that he would have approved of my book as it is now presented, and that others will find it helpful, and to their liking.

D.M.S.
November 1996

232

Notes

Chapter 1. Retrospect (pages 5- 17)

1. George Judge was Hon.Sec. of Kent BKA from the founding of the Assoc. (largely on his initiative) in 1916, until 1946, when he became President. "Bee Craft" was also his idea, and started as a duplicated sheet in 1917 and as a monthly journal in February 1919. Judge offered the facilities of the journal to the Surrey BKA in 1920 and to others subsequently, so that by 1925 "Bee Craft" was serving five of the largest BK Assocs. in the country, and had a circulation of 2,500 monthly. By 1937, when H.J. Wadey took office as Editor, the circulation was 3,800 copies monthly, serving ten BK Assocs., and by 1947 was 12,000 copies, serving 20 BK Assocs. In 1947 it was accepted as the official organ of the BBKA. Judge consistently urged the adoption of federated regional grouping to the British Beekeeping organisations, and made proposals to this effect to the BBKA in 1919 and again in 1938. It was on his initiative that the South Eastern Federation of BK Assocs., comprising Kent, Surrey and Sussex, was founded in 1932. By 1946 there were eight such federations, and only twelve counties not so federated in the regions.
2. Arthur Dines kept bees for many years in the Dartford area of Kent, and near Canterbury after he retired. His active participation in beekeeping affairs, in the County Association and Branch, the BBKA, the National Honey Show, and with Bee Craft, spans forty years or more. He was President of the BBKA (twice), Assistant Editor and then Editor of Bee Craft (following H.J. Wadey) until 1981, and contributed a monthly series "For Beginners" in Bee Craft that must rank as a major contribution to better beekeeping.
3. My acquaintance with Barnes sprang from being sent to tell him of a swarm, and watching him take it. It was the first of many I helped him with. He gave me a Woodman smoker that I used for fifty years. I kept in touch with Barnes until he died, and visited him occasionally. He was a good friend to me.
4. Leonard Illingworth (1882-1954) was Secretary of the Apis Club from 1936 onwards, founder and Secretary of Bee Diseases

Insurance Ltd., and President of the BBKA at the time of his death. In its early years "Bee World" was printed locally, and wrapped and posted to subscribers by Illingworth and his sister. His sixty colonies were in Langstroth hives, with a deep and a shallow (Langstroth) as a brood chamber. They were all in one apiary, half a mile from where I now live, in Foxton, near Cambridge. My next door neighbour, Wally Stock, used to help Illingworth with his bees. Wally died in 1975, aged 84, having lived in Foxton all his life. He had kept bees himself, as his father had done before him. Wally told me that there were more than 100 colonies in Foxton in the 30's and 40's, which got highly satisfactory crops of honey from white clover and sainfoin. Illingworth (and Wally Stock too) gave up beekeeping in the early 1950's, when crop spraying killed bees to a disastrous degree and both clover and sainfoin ceased to be grown in the vicinity. The area is still predominantly farmland, but there is little to support bees in Foxton today.

5. Gauntlet Thomas lived and kept 200 or more colonies at Exning, near Newmarket. The sainfoin was grown locally for hay for racehorses in the Newmarket training stables. Thomas also took colonies into Breckland (where Madoc later kept hundreds of colonies) for the heather.

6. Samuel Simmins' book, "A Modern Bee Farm", ran into four editions, the first in 1888 and the last in 1928. Secondhand copies of the later editions can be found without difficulty. It is a very interesting book, well worth reading. His advice on colony management is generally sound and still relevant; his ideas on hives (despite his advocacy and use of the 16 x 10 frame) far less so. Simmins ran what was perhaps the first commercial beekeeping business in the UK, with more than two hundred colonies. He had a covered apiary (three long bee houses, with extracting, bottling and storage room) that could accommodate one hundred and fifty colonies. He was an early importer of queen bees from southern Europe, and bred bees for sale on a considerable scale. His booklet on Queen Rearing (1894) is sound, and his comments on the various races and strains of bees, and their crosses, are interesting. Simmins lived at Seaford, and later at Heathfield, in Sussex, and had his bees for many years in the chalk downland country behind Seaford and Eastbourne, where Sturges later kept bees on a commercial scale, and John Hunt and I had bees in the 50's. The Forestry Commission's Friston Forest now covers part of this area. I met Simmins only once, in the summer of 1928, when I went with some other beekeepers to his apiary at Heathfield. It was considered a great privilege, as I recall. I was barely fourteen years old then; he was past seventy, and held in very high regard. The last edition of his book "A Modern Bee Farm" (of 530 pages)

234

was published in that year. I bought a copy (for 3/6d) which I still have. I also have a copy of the 1893 edition. Simmins died in 1940, aged 83.

7. The summers of 1933 and 1934 were long, hot, and very dry, in Suffolk (and, I think, elsewhere also). The total rainfall recorded at the local weather station, for the two years combined, barely amounted to the long term annual average.

8. Change in the rural/farming scene since the 1930's has been quite dramatic, not least in the effect on beekeeping and beekeepers. I have written about this elsewhere, (See Appendix 1) but here I would repeat that the changes in farm practice of particular significance for beekeeping have been the control of weeds, pests, and diseases by spraying, and the cutting and harvesting of forage crops at an earlier stage of growth than hitherto, i.e. before flowering. In contrast to the 1930's or 40's, many rural areas, at least in the south and east of England, and except for oil seed rape, are no longer much good for honey production.

9. Until a few years ago John Hunt had about 20 colonies of bees in a variety of hives (Dadant, Langstroth, Jumbo, Smith, and 14x12) by choice, for interest and comparison. He also had the local Branch apiary (in Nationals). Now. like me. more than 80 years old, he has reduced colony numbers to four (Smith and 14x12) in a bee house in the garden.

10. We got to know R.P.Sims quite well. He (and his father before him) used a hive similar to that used by R.W.Wilson, a 10-frame Smith. He ran a 400 colony honey production unit, and following his move to the Canterbury area he grew dessert apples on a commercial scale, and had his mininucs in the home orchard.

11. Seed crops of all kinds, both farm and garden seeds, and notably Kent wild white clover, have long been grown in Romney Marsh, for three good reasons, viz: soil fertility, favourable weather to ripen seed, and opportunity to isolate. Oil seed rape is now also grown there in quantity, but, alas, wild white clover for seed is no longer grown.

12. Romney Marsh dykes are 20 feet (6 metres) wide, with controlled water levels.

13. Red clover, left for seed after an earlier cut for hay, was well worth taking bees to for a late and sometimes copious flow. John Hunt and I once found our bees working bell heather (erica cinerea) - more than an acre of it - high on the chalk downs, when we set off in the Land Rover to find the source of the reddish honey that the bees were storing. A few small pockets of acid soil overlie the chalk at the highest levels, which permit the presence of ericaceous plants. It was in this part of Sussex that A.M.Sturges built up a substantial honey production business of 500 Langstroth colonies

in the `30`s, in apiaries with about 20 colonies in each, and continued to breed and sell queens from 500 mating nucs, as he had done earlier in Cheshire. There, in the `20`s, he had imported, bred, and sold Italian queens, and wrote a book (in 1924) and several shorter pieces. I never met Sturges, but Wadey knew him well and used to tell me about him and his outfit. Wadey writes about him and his partner Leslie Soden in "The Behaviour of Bees and Beekeepers". Sturges developed what Wadey calls "a huge `over and over` extractor (that is an extractor with a horizontal spindle) that held 56 combs. .It was the forerunner and prototype of Abbott`s MG Parallel Radial extractor, which I (and Colin Weightman too) have and use. Ours hold 21 combs, of any size.

14. Beowulf Cooper was a colleague of mine in MAFF, at that time. He was a pioneer in demonstrating to farmers, and to agricultural advisers, the value of honeybees as pollinators of farm crops, as Fred Richards was in Norfolk.

15. In one apiary of 12 colonies the colony yields ranged between 187 lbs. and 236 lbs. (average 213), and in the other apiary, also of 12 colonies, a mile away, the colony yields were between 159 lbs. and 217 lbs. (average 191). The overall average, for the 24 colonies, excluding heather honey, was 203 lbs., viz: 38 lbs. fruit blossom honey, plus 165 lbs. in Devon. The three highest colony yields, of 236, 227 and 220 lbs. were all from colonies headed by sister queens, the daughters of a Buckfast breeder queen. In "The Border Bees" Colin Weightman records yields in the same year, 1955, of 208 lbs, 204 lbs, and 186 lbs of clover honey in his home apiary in the Tyne Valley.

16. In 1953, at apple blossom time, Fred Richards, with other CBI's, visited two of the apiaries that John Hunt, J.L.Liebenrood and I ran in Kent. Fred Richards had developed a system of management akin to that used by Rauchfuss (see the chapter on Management Systems) which we tested. Both in Devon and in Norfolk his emphasis on the use of a suitable strain of bee and of simple and effective management methods have influenced many beekeepers, including myself. His experience with bees is lifelong, and immense.

17. The Exmoor site I found and used for some years was near one of the sites used by Alec Gale, of Marlborough. I met him on several occasions. He was always helpful and interesting. His experience of bees was vast, and his attitude open minded and essentially practical. Having tried others (on a considerable scale) he used 13 BS frame hives. On Exmoor, the bees could be found working wood sage (teucrium scorodonia), rosebay willow herb (epilobium angustifolium), thistle (cirsium sp.), and bramble (rubus fruticosus) at times, although the principal source of honey was the ling

236

(calluna). The resultant honey had the true heather honey character (thixotropy, aroma and "tang"), but it was somewhat lighter in colour and more ready to granulate than the Dartmoor honey or than that which I later had from the Northumberland moors. New Forest heather honey was very similar to that from Exmoor. I expect it still is.

18. The period of "at least 15 weeks" was from 23 November 1962 until 8th March 1963. A few bees flew (over snow) on Christmas Eve. Tests for nosema, which seemed the likely cause of the failure of colonies to build up in the spring, were negative.

19. Colin Weightman records (in "Beekeeping in the Shilford Apiaries", 1954) a ten year colony average (1942-52) of 54 lbs. (including heather honey) from 120 colonies. His colony average for the following ten years (which included 1955) was much the same.

20. The Northern Federation of BKAs put on a splendid display at the Show, and the prize classes were uncommonly well supported.

21. Ernie Pope kept about 20 colonies in Yarrow hives for nearly forty years until very recently, when age and ill health forced a reduction in colony numbers to six, and finally to none, The Yarrow hive is, in effect, a 9-comb Smith, developed at Auchincruive, constructed of ½ inch exterior grade plywood, essentially for double brood box working (18 combs) and a single box move to the heather. Six or eight of Ernie's colonies were kept in a bee house in his garden. Regrettably, both the beehouse and the bee colonies are no longer there. One advantage of a bee house, valuable in an urban area, is that it conceals from neighbours, and from vandals, the fact that bees are kept, and thus reduces both damage and complaints.

Chapter 2. Bees and Queens (pages 19-42)

1. Colin Weightman's book, "The Border Bees", published in 1961, is unusually informative, about bees, beekeepers, colony yields, and colony management in a heather district. He has been a regular contributor to The British Bee Journal for many years, writing "from the North", and lately "Combings", in that Journal. His hospitality to visiting beekeepers is legendary, he has friends in many countries, and has accompanied Brother Adam on his visits to Greece and elsewhere.

2. Hasting's Caucasians stored honey in a minimum of comb space, wintered well, and economically, without much feeding, made powerful colonies for the heather, and did well there. They were more prolific and continued breeding later in the year than the local bees, but were not profligate. They were uncommonly quick to get rid of drones after the first frost. They were easy to handle,

and to manage. The queens were easily found. Crosses with the local bees were very good, no matter which way the cross was made.

3. The BIBBA publications are now numerous, and all useful. Some provide information not readily available elsewhere, e.g. on mininuc construction and use.

4. Pettigrew`s book "The Handy Book of Bees", published in 1870, was perhaps the most popular book on bees for some years thereafter. It advocated skep beekeeping, in a skep of much more than usual size, the so-called bushel skep, used by his father for many years, and subsequently by himself. He thought wood an unsuitable material for beehives. William Cobbett, writing in "Cottage Economy" fifty years earlier, (1821) was of the same opinion - wood was too cold for bees. The Pettigrews were based and had their bees at Carluke, in southern Scotland, Manley discovered bees in Northamptonshire kept in Pettigrew or bushel skeps, and says (in "Honey Farming" page 16) that they gave about three times as much honey as the usual small ones.

5. Beowulf Cooper and I discussed bees and beekeeping on numerous occasions, both before VBBA (now BIBBA) was formed and during the years that followed. He ran 30 or more colonies in out-apiaries, as I did, and had the same aim, namely maximum returns with minimum labour, and for the same reason, namely limited spare time. We found a large measure of agreement, both about bees - good wintering, disinclination to swarm, docility, behaviour on combs, industry, etc., and about colony management - wintering with ample stores and little or no spring feeding, no special steps to discourage swarming other than the provision in good time of ample room, the importance of near perfect brood combs, all new brood combs drawn above the brood chamber, etc. But we disagreed about bottom bee space and glass quilts, which I dislike and don't use, and, more importantly, about brood chamber size. I share his view that the way ahead is through breeding better bees, better suited to the locality in which they are kept, but I could not, and cannot, accept, as a desirable aim, bees whose brood chamber requirements are met by a single brood box of eleven 14 x 8½ inch combs. I consider a brood chamber of that size inadequate for really productive bees, even of near native strains.

6. Steve Taber says that in California, beekeepers make over half their income from pollination, which opens the door for queen breeders to develop strains of bees more inclined to collect pollen or to visit certain plants for pollination purposes. He cites proof that it can be done. (Steve Taber, Breeding Super Bees, The A I Root Co., 1987). Brother Adam has pointed out that some strains of bees

are useless at the heather, and the ability to work ling heather effectively has to be bred into the bees.

7. In his book "Rearing Queen Honey Bees" Roger Morse states his dislike of Caucasian bees because they gather and use so much propolis, with the result that fingers get stuck up with it. I agree with him that bees that propolise excessively make comb handling sticky, and often difficult. In my experience, while strains of bee vary greatly in the extent to which they use propolis in the hive, and the manner in which they do so, the presence or absence of the material for collection is a potent factor. Poplar trees are a menace in this respect. Brother Adam says that the tendency to build brace comb is fairly easy to breed out, whereas that for propolising can only be eradicated at the cost of endless trouble. ("In Search of the best strains of bees" page 179)

8. Dr. C.C.Miller died in 1920, aged 90. He had kept bees for sixty years, and the production of comb honey was his sole livelihood from 1878 onwards until his death. In 1913, at the age of 83, he broke all records of production per colony of comb honey sections. For fifty years he made regular contributions to the American bee journals, and his writings, were, and in some respects still are, very influential. His book, "Fifty Years Among the Bees", published in 1911, was reprinted in 1980. It provides a fascinating picture of early commercial beekeeping, and is written with a delightful style and remarkable clarity.

9. Peter Kemble considers Cloake's method to be particularly suitable for rearing queens early in the season.

10. Vince Cook, the MAFF National Beekeeping Specialist, and former New Zealand Beekeeping Adviser, in the BBJ, Oct. 1980. His system of queen rearing is set out in his book, "Queen Rearing Simplified" (B.B.Pubs.1986). The description is admirably clear and the operations are very well illustrated.

11. Wadey's two books, "The Bee Craftsman", and "The Behaviour of Bees (and Beekeepers)", both published by Bee Craft, are still available, much underrated, and well worth having and reading. They are full of good sense, and amusing too. I knew Jim Wadey well in the years just after the war, when I lived only five miles from his house, and I spent many a summer evening with him and his bees, and attended more than a few of his winter evening talks to beekeepers. Wadey was Editor of Bee Craft for many years. As a lecturer at bee meetings he had no equal, and was in great demand.. He always spoke from a single post-card "aide memoire" (without slide projection) and urged others to do likewise.

12. In Bee Craft, January 1987, Aebi reports similar successful use of pulled virgins.

13.The Worth cage (see Fig 18) has a facility for access to the queen by worker bees through an excluder slot and the delayed release of the queen until a small piece of candy is consumed. I cover the open end with a small piece of paper, secured by a rubber band, as I do with the Butler cage. The Fileul cage was very similar to the Worth cage, but seems no longer to be available. This principle of delayed release was inherent in the Manley cage (see Fig 20 and Manley's book "Honey Farming" pp. 200-208) in which the queen is caged on the comb. In my experience queen introduction in the Manley cage rarely fails and I still have in use the six Manley cages that I made in 1947. But Manley cages cannot be bought, as far as I know, and are not easy to make. All these cages are intended for the introduction of queens unaccompanied by worker bees, which I (and many others) think best. I tend to use the Butler cage as a rule, but I can't say that I have found any significant difference in the effectiveness of one or the other. All three are very satisfactory.

14.In "Bee Craft" May 1968. R.O.B.Manley died in 1977, aged 90. His first book, "Honey Production in the British Isles", published in 1936, was an eye-opener to those who read it, as I did in 1937. He had given up farming in about 1926 to devote his whole time to beekeeping and gain his livelihood thereby, as he continued to do throughout his long life. His second book, "Honey Farming", published in 1946, became a classic, and, for many, the serious beekeeper's bible. He followed this with "Beekeeping in Britain" in 1948. Manley's contributions to Bee Craft were frequent, always very much to the point, and deserving of attention. Most of them are as relevant today as they were when they were written.

Chapter 3. Handling Bees (pages 47-55)

1. Beowulf Cooper also much disliked what he called "followiness" in bees, and sought to eliminate it in bee breeding.
2. I change the sequence of comb faces in the brood box only with good reason. Brood combs are rarely or never completely flat and of uniform thickness (although, with care, they will come near to it) and I seek to maintain the maximum usefulness of all the comb faces, and uniform spacing between combs. Lack of attention to the sequence will lead to comb spacing that is far from uniform, and to drone cells where the spacing is wide and bald patches where it is narrow.
3. Coverdale Management Precepts.
4. I like to have young queens mated and laying in nuclei as soon as possible each year, usually by the end of May. It pays handsomely

to rear a few early queens from a selected breeder queen. One can re-queen colonies by adding the whole nucleus.

5. Read what Dr.A.L.Gregg has to say about bee stings in his book, "The Philosophy and Practice of Beekeeping", (Bee Craft, 1945.) A most interesting, indeed fascinating, book that gives much food for thought. Dr.H.R.C.Riches has a Chapter on bee stings in his book, "Beekeeping", (Foyles, 1976.)

6. R O B Manley wrote: (In the BBJ August 1962) "If you work among bees all day, you really need gloves. If you don't use them, your hands get into a ghastly mess Of course you can't handle queens with gloves on, or with propolis covered fingers, so wear gloves and take them off to handle a queen. Bees sting gloves rather more than they sting bare hands, but if washed in weak IZAL solution they will not sting either very much."

Chapter 4. Hives and Equipment (pages 57-86)

1. David Little's report of the BBKA Survey of Beekeeping in England 1981-85 (in BBKA News June 1986) includes percentage figures for the various hives in use. Based on more than 1,000 replies to a questionnaire, the figures are: National 53%, WBC 23%, Others 24% (Langstroth 2%, MD 8%, Commercial 9%, Smith 4%, Others, including B.Deep 1%). It would be interesting to include Scotland and Wales.

2. The metal spacers offered (until recently) by the A.I.Root Co. are much the best form of "clip" or spacer for converting 7/8ths inch wide side bars to Hoffman spacing. I have many in use as spacers on my follower boards (See Fig 4). I am hopeful that these spacers may soon be made in England and made available to British beekeepers.

3. Users would presumably consider British Deep (14 x 12 ins) and British Commercial (16 x 10 ins) hives to be of adequate size for single box working. Perhaps they are; certainly both are so used. Relative to a single National or Smith brood box, the capacities (total surface area of frames) of the various brood chambers in use in the UK are as follows:

Single deep	No.of frames	Relativity
National or Smith	11	100
Langstroth	10	123
British Commercial	11	134
British Deep	11	141
Langstroth Jumbo	10	151
Modified Dadant	11	166
12-frame Dadant	12	182

deep + shallow		
National or Smith	22	165
Langstroth	20	197
double deep		
National or Smith	22	200
National or Smith (2x10)	20	182
Langstroth	20	246

4. Wadey lived at Crowborough, in Sussex, his business was tailoring, and he kept bees on a semi-commercial scale in and around Ashdown Forest, nearby, where there is heather (both bell and ling) in some areas. He had his bees in double brood chamber Langstroth hives, with the upper box set back an inch or so from the lower one, to give a full width entrance between the boxes, and kept the bottom entrance small with an entrance block I know of no-one, other than he, who would hold a queen bee between his lips if he needed to have her secure for a time, which I saw him do more than once..

5. As Athole Kirkwood found, when he had more than 1,000 colonies in Smith hives, equipment for large scale operation is designed to handle Langstroth frames, and both costs less and handles more when so used than when it is adapted for BS frames. He consequently changed to Langstroths and now prefers the Langstroth hive as a honey production unit in a heather district. (See Mr Kirkwood's article on the subject in The Beekeeper's Annual 1986)

6. In Bee Craft October 1938, WHH (the initials of a very well known name in beekeeping) wrote "... we must have more than is provided by ten BS frames. In fact twenty BS frames is not too much ... It has been my experience that swarming may be more easily controlled where two boxes of smaller frames are in use than when one box of big ones is used. The Langstroth hive is not a single chamber hive. It is hopelessly small, and its ten frames are inadequate for a normal queen's production. Twenty of them are not too much for a good queen and a reasonable food reserve. Here we have a really adequate brood nest, substantially larger than the MD hive, and also have the adaptability of the two chamber hive to many forms of swarm control manipulation."

7. The use of two ten-frame Langstroth hive bodies as the brood chamber is standard practice in many parts of the world. But the ten-frame boxes usually contain only nine frames, not ten, in the upper one, and eight frames and a frame feeder in the lower one. It is interesting to note that these seventeen Langstroth frames have the same comb area as twenty-two BS combs, such as I use in

242

my Smith hives, and many beekeepers in the UK use in National hives.

8. My Smith boxes have precisely the same capacity, number of combs and frame size, as Nationals, and all the operations and management practices that I describe are equally applicable to colonies in National hives. I prefer the Smith hive because of its top bee space, short lugged frames, and truly single wall construction; others, and evidently an overwhelming majority in England, prefer the bottom bee space and long lugs that National hives provide. But the choice, as between National and Smith, has little or no effect on operations and management. What I do, with Smith hives, can equally be done with Nationals, (or, for that matter, with Langstroths).

9. In his book "Bee-keeping with twenty hives" (A Bee Craft book, 1952) A L Sandeman Allen devotes a chapter to hives and frames, in which the merits of alternatives are usefully discussed.

10. A.M.Dines, "For Beginners", Bee Craft, November 1969. Arthur Dines' series in Bee Craft "For Beginners", which ran for many years, conveyed sound, sensible, advice and comment on a wide range of beekeeping topics. Re-reading the series, and making notes from them, is a rewarding winter study.

11. W.H.H., writing in Bee Craft, October 1938, says: "Top spacing, used with Hoffman frames, results in an immense saving of time, as there is then no need to lever the bottom edge of every super away from the top bars of the frames beneath. It is most regrettable that the chance to make this change was missed when the National hive was introduced. When a feature is a definite improvement we should face the fact and make the cut."

12. Dr.C.C.Miller, op cit. The BBKA has a useful leaflet (HA5) on frame spacing, which deals at some length with spacing by screw eyes.

13. M.A.Alber, paper to 23rd International Apiculture Congress, 1971,; R.J.Pearce et al (1963); and Johansson and Johansson in Gleanings 1985.

14. See also Note 2 above.

15. In the American Bee Journal, December 1988, Hugh Maxwell says: "Most beekeepers .. start with the second frame in; the first one is stuck most often. It is so hard to remove that even if it comes out in one piece, the bees have become alerted and are scenting and whirring an alarm."

16. op cit.

17. The A.I.Root Co's Catalog offers the slatted rack - in Langstroth size, of course.

18. DMS, article in The Scottish Beekeeper, April 1975, and Graham Burtt in the following issue.

19. I am told that lockslides are of Australian origin. Steele & Brodie and Thorne both supply them and illustrate them in their catalogues.
20. In "A Manual of Beekeeping" para.874, page 226.
21. I have tried several of the new types of bee escapes. Those fixed under a central hole in a clearer board have the merit of having no moving parts. They work well, and, like the Porter escape, leave the supers quite empty of bees. Used as they should be, and not left on hives for days or weeks, they are not much propolised, but if they do get gummed up they can be restored, as Porter escapes can, by boiling them for a minute or two in a strong solution of washing soda. But they still require supers to be left overnight for collection the following day. In out-apiaries and on the rape we need a clearer board that will empty supers of bees in two or three hours. The new multiple cone escape with a deep rim is said to do so with minimal stress to the bees. It works well, and quickly, but not quickly enough for me. So I have at last bought, and now use a petrol driven blower to get bees out of supers at oil seed rape time. We tried out, and then bought, a blower designed and sold for blowing and picking up leaves and debris in parks and large gardens. It blows bees out of supers, taken off and stood on edge, very quickly and without harm or stress to the bees. We can then load the cleared supers at once, and start extracting the honey from them an hour or two later, and return the extracted supers to colonies next day. This machine cost much less than any purpose designed bee blower, and can also be used for the purpose for which it was designed. I have an even less costly garden leaf blower and collector which blows bees out of supers almost equally well, but it is powered by an electric motor and can be operated only within cable length of an electric power point, so is useless in out apiaries. This machine, which I bought for garden use, could suit beekeepers with hives in their garden as a bee blower.
22. Clearer boards of this type vary in design. Some (among them the so-called New Zealand type) have an escape hole in one or more of the four corners and a narrow (screened) exit below, through which few bees find their way back. Boards fitted with cone escapes are in use in USA.
23. A Manley cage (see Fig 20) consists of a wooden frame of half inch thick material about 3 ins. x 2 ins. overall, covered on one side with wire cloth. On the other side a collar of perforated zinc is fixed, either by cutting and folding the perforated zinc itself or by soldering it on to four tin lugs. Two holes are bored (and scorched to ensure a smooth passage) through the frame from outside to inside, and a piece of queen excluder is fixed on the inside of one of these. A crate staple, such as are used to fasten floors to hives,

driven in to the frame of the cage, spaces the cage from the adjacent comb and prevents the cage from falling. In use, a very small candy plug is put into the hole with the queen excluder and a larger plug put into the other. (See "Honey Farming" pp 205-208) The cage is very suitable for use in out-apiaries as further attention is not required until the next routine visit, a week or more later, when the cage is removed and a check made that the queen has been accepted. In my experience queen introduction by the Manley cage very rarely fails.

Chapter 5. Colony Records (pages 87-97)

1. The BIBBA record cards are excellent, for their purpose, which is primarily the evaluation of a queen as a potential breeder. The colony record that I keep is primarily a management tool, and only in part, in review, an aid to evaluating the queen.
2. K.K.Clark, "Beekeeping", Penguin Handbook, 1951. Regrettably long out of print, but used copies can be found and bought. An excellent little book, full of sound observation, comment, and advice, still very relevant.
3. Dr.C.C.Miller, op cit.
4. DMS, articles in The British Bee Journal, September 1955 and November 1972, and in The Scottish Beekeeper, September 1977.
5. To illustrate what I write in my record book at the end of each year's records, usually in mid October, this is what my record book says about 1988, a bit below average sort of year. The following year, 1989, was a very good year; in fact one of only three years in the past thirty-five years in which the colony average for a forty (or more) colony outfit has exceeded one hundred pounds. Very warm spells in the last two weeks of May (OSR time) and of July (borage) produced exceptional results, a colony average of 118 lbs., and a total crop of 5660 lbs., including 550 saleable round sections. 8 colonies exceeded 200 lbs. But in 1988 the weather at oil seed rape time was not very good, and likewise in July, when the borage was in flower. The record book for 1988 says: Autumn count 46. 4 died over winter. Woodpecker damage severe. Total crop 3140 lbs., including 270 round sections (150 lbs.) Colony average 68 lbs. No colonies moved to rape; 4 moved to borage. Highest colony yield 94 lbs. from a 14 x 12 colony managed a la Zimmer. Colony average of the 8 Zimmer colonies 81 lbs.; increased the number of these to 12. Main sources of nectar, oil seed rape, field beans, borage, willow herb, and bramble. Re-queened 22 colonies. 54 colonies to overwinter, including 6 nucs.
6. Op cit.
7. See Note 15 to Chapter 1

8. "Winter count" was also Manley's basis of assessment. On a "spring count" basis - which most beekeepers seem to use - averages can be very different (and always greater).

Chapter 6. Colony Management (pages 99-120)

1. H.J.Wadey, in "The Bee Craftsman".
2. Sue Hubbell, in "A Book of Bees", New York, 1988, says: "When I first began with bees the great diversity of passionately held opinion bewildered me, but now that I have kept bees in widely different locations I think I understand. Frost comes earlier in some places than in others. Spring comes late. Rainfall is not the same. The soils and the flowering plants they support are unlike. Most people who keep bees have only a few hives and keep them all in one place. They find it difficult to understand how practices they have found successful do not work for others. But I have learned that I must treat the bees in one yard quite differently from those even thirty miles away. So it is no wonder that what works well for a writer based in one place may not work at all for one based in another. The thing to do, I discovered, was to learn from the bees themselves."
3. DMS, lecture notes, revised for publication in Bee Craft, October 1954.
4. E.R.Bent, "Swarm Control Survey", (Gale & Polden, 1946.) Regrettably long out of print. Ref. also E.B.Wedmore, "A Manual of Beekeeping", 1945; and Frank C.Pellett, "A Living from Bees", 1945.
5. "Dividing Over-wintered Colonies for Increased Honey Production", Ed. Braun, Dominion Experimental Farm, Brandon, Manitoba, Canada, Dec. 1945; also Experimental Farms Service, Canadian Department of Agriculture, Publication No.774, 1945, and Instruction Sheet, March 1951.
6. In Chapter 7 "Management Systems".
7. S.C.Jay, "Drifting of honeybees in commercial apiaries", Journal of Apicultural Research.
8. op cit.
9. My keynote address "Changes in farming practice in England and Wales and their implications for beekeeping", given to the BBKA Spring Convention at Stoneleigh in April 1983, and published in "Bee World" later that year, deals with this subject more fully, and covers the fifty year period 1932 to 1982. It forms Appendix 1.
10. H.Mace, in "The Complete Handbook of Bee-Keeping", Revised Edn. 1976. The picture of 30 beehives at the heather on page 128 of that book is of my hives, above Blanchland, in 1964.
11. See Note 6, Chap.12.

12. R.O.B.Manley, in Bee Craft, January 1938, says: "It has been suggested that by continual breeding from strains of bees that show a strong tendency to supersedure of their queens we may arrive at the desired end, but personally I am dubious. I can remember a dozen poor queens produced by supersedure for one really good one. I do not remember a single instance of a really outstanding queen that had been produced in this way. Stocks that have a strong tendency to supersedure are rather apt to turn up queenless in spring through attempting supersedure late and failing to have their queen mated."
13. Roger Falk, in "The Business of Management" (1961) says: "Management is largely a matter of decision, and decisions cannot be properly taken unless the mind is clear about objectives and priorities. Clarity of mind calls for concentration, which is in itself a vital discipline."

Chapter 7. Management Systems (pages 129-140)

1. R.W.Wilson, in "Beekeeping", Devon BKA, May 1948.
2. "Researches on methods of management of bees: Final Report", March 1952, by E.B.Wedmore. Published by B.R.A. April 1952. Investigations conducted for the BBKA Research Committee.
3. E.B.Wedmore, op cit
4. J.Ashton, "Reverse Snelgroving for Heather Honey Production", Kirkley Hall, Feb. 1972.
5. In "A Manual of Beekeeping" para 1448.
6. Reported in The Scottish Beekeeper, April 1965. Ian Maxwell, NDB, Lecturer in Horticulture and Beekeeping at the West of Scotland College of Agriculture, Auchincruive, still uses the method.
7. Illustrated at Fig 17
8. Bert Mason, of St Cyrus, Montrose, some 30 miles south of Aberdeen, runs 800 to 1000 colonies in Smith hives, single handed except for help with moving and at extracting time. He moves colonies to oil seed rape, to raspberry, to bell heather, and all the colonies to the ling heather. His system largely eliminates swarming, increases colony numbers, and results in highly satisfactory crops of honey. His operations are, of course, timed to suit local conditions; they could be expected to be a week or two earlier in southern England.
9. In 1985 Brother Adam drew my attention to the system adopted by Raymond Zimmer, at Colmar in southern Alsace, 50km north of Basel, where large areas of oil seed rape (colza) are grown. Zimmer uses Buckfast bees in a 12-frame Buckfast Dadant hive, to which he unites a 12-frame Langstroth colony with a young queen in

September. (Later putting the Langstroth box below the Dadant). He removes the Langstroth box in March, and takes his Dadant colonies to the oil seed rape, and later to acacia, to chestnut and lime, and to spruce fir. Queens are reared and new colonies built up with the young queens in the Langstroth boxes for autumn uniting. His takes of honey are very impressive. Vide Raymond Zimmer, "L'Abeille Buckfast en Question(s)", Zimmer 68-Horburg-Wihr, 1985. (FRF 50.00). In Bee Craft, August 1985, Jim Holland refers to Zimmer's system and suggests using British Commercial and National boxes. My own use of double brood box Smith colonies in Essex has led me to adopt the system with a combination of Smith and British Deep (short lugged/top bee space).

Chapter 8. Swarm Control (pages 141-155)

1. op cit. Part Two of this book, An Evaluation of the Races and Crosses, is fascinating, and uniquely informative.
2. Manley, in Bee Craft, January 1938, says: "The greatest of all factors in swarm elimination is strain. Some strains swarm excessively, some very little, though none is exempt from swarming. There are strains of all varieties of bees that swarm less than other strains of the same varieties, and it is by breeding from drones and queens of such strains that we can reduce swarming to profitable limits."
3. In Gleanings in Bee Culture.
4. Dr.A.L.Gregg, op cit.
5. Taranov swarming first came to my notice in 1952. In "Bee World", under the heading "The Artificial Separation of the Swarming Bees from the Parent Colony", C.F. Taranov wrote at some length, describing the method developed from 1945 onwards. He said "We believe that by using our method, one gets the basic mass of bees that would have swarmed if left alone ... there are some differences, namely 1. in the artificial swarm there are bees of only 1 to 3 days old, too young to have taken part in a natural swarm, and 2. many flying bees from the parent colony and from others join a natural swarm - indeed if it is long in the air, or if it clusters for some time in the apiary, it may gain 1 to 2 kg. of these recruits - whereas in the artificial swarm 80 to 90% are the idle bees, which are the basis of either kind of swarm. We made a special board consisting of two boards 25 to 30 cm wide and 50 cm long, nailed together at one end, and at the other there are two supports, the height of the alighting board ... we put a sack or cloth behind the board and right up to it ... when we shake the first comb we aim to have some of the bees falling upon the cloth and some upon the board; the other combs

are shaken gradually further and further from the board. The bees walk up the board immediately or in a very short time. In warm weather it is more satisfactory to make the artificial swarm in the evening or to shade the board; the bees then cluster more closely. This kind of swarm should be hived through the entrance; the bees start work sooner. If there is no nectar flow, they should be fed. When hived the same evening or the next day, of course with the queen, they remain there, even if their hive is next to the parent one, and work energetically." John Hunt and I made a board or two in 1952 or 1953 and followed the instructions, and I have used a Taranov board from time to time ever since.

6. Richard Taylor, "The New Comb Honey Book", 1981. A book full of interest and good sense, delightfully written. So too are Dr. Taylor's other books "The How-To-Do-It Book of Beekeeping" and "The Joys of Beekeeping", and his "Bee Talk" series in Gleanings in Bee Culture.

7. Dr.A.L.Gregg, op cit.

Chapter 9. Oil Seed Rape (pages 157-171)

1. In the one million acres of oil seed rape that we now have in the UK, the proportion of spring sown rape, although still small, is greater than it was, so that fields of rape in flower in summer are not uncommon, much to the advantage of beekeepers.

2. F.N.Howes, "Plants and Beekeeping", Faber, 1945. John Free has added a section on oil seed rape to the new and revised edition, Faber, 1979.

3. Systems of management are now being adopted that put into the rape really powerful colonies headed by young queens. (See Note 9 to Chapter 7).

4. F.C.Pellett, op cit.

5. The need to spray oil seed rape may be further reduced by the development of seed treatments. The incorporation in pesticides of substances repellent to honeybees for long enough for the pesticides to be effective, but fading to allow foraging to be resumed safely, is another likely development, according to Dr. J.A. Pickett (Gooding Memorial Lecture, 1985, CABk). Where my bees are kept, in Essex, the farmers use Fastac, which incorporates this feature and is harmless to bees.

6. On a larger scale, an arrangement for blowing warm air below several stacks of supers is useful. The temperature should be thermostatically controlled at about 35 degrees C (94 degrees F).

7. In 1989 I put into use, at the oil seed rape, the ekes that I had used in past years for the production of comb honey at the heather. I did so simply because I had run out of supers in an exceptional honey

flow, but I now do so routinely. These ekes are made to sit inside a brood box (two in a box) with a bee space all round, maintained by staples or studs. A starter of thin foundation on which the bees build comb is fixed to each of the nine bars in each eke. See Fig 27. Fully drawn and filled, the two ekes will together hold about 45 lbs. (20kg) of honey. Used on the rape the filled ekes are stored (granulated) until September, when the comb is cut from the bars, chopped up, and melted. If time presses at next year`s rape flow, as it usually does, the ekes go back onto colonies without starters, as some of my supers also do. There is usually sufficient of a starter left by the knife to get the bees building comb much as one would like.

8. A Strainaway warming and straining outfit which I bought in 1992. It does everything I need easily and quickly. The liquid honey is strained by vacuum pump through a chosen mesh screen and is beautifully clear and clean. I also have an American (Kelley) cappings melter and wax separator which melts granulated combs in a water jacket at a controlled temperature. We commonly use both.

9. Reported in The Scottish Beekeeper, June 1985.

10. A similar technique, applicable to comb honey sections, is described by Carl Killion, in "Honey in the Comb", 1951.

11. Whether or not there is a detrimental effect is disputed. It may depend upon the design of trap. A facility for removing the trapping screen without disturbing the colony is clearly desirable.

12. I have seen OAC pollen traps in use in Spain, where newly collected pollen can readily be bought and is exported. Luddington use a pollen trap of their own design and will supply details on request. I have one of their traps and also one of another pattern on sale in the UK. Both work well, but I prefer the OAC trap; it collects more pollen. I was told of a beekeeper in British Colombia who has more than 2,000 OAC traps in use through May and June and traps about 15 lbs of pollen per colony.

13. T.S.K. and M.P. Johansson, "Some Important Operations in Bee Management", IBRA, 1978.

Chapter 10. Out-apiaries (pages 173-181)

1. Peter Beckley, "Keeping Bees", Pelham 1977.

2. Wadey devotes a chapter to out apiaries in his book "The Bee Craftsman", (A Bee Craft book, 1943).

3. The contents of my bee box are: smoker, with attached Root type hive tool, J-type hive tool and spare Root type hive tool, smoker fuel, matches, record book with pencil attached, knife, scissors (for clipping queens), small hammer, small saw, secateurs, screwdriver,

pliers, box of assorted small nails and screws, a clean, empty, matchbox or two (for temporarily holding a queen), a few queen cages (Manley, Butler and Worth), tissue paper (for lighting the smoker), some propolis scrapings (for adding to the smoker fuel), some pieces of foam plastic (for temporarily closing hive entrances), some plasticine (for stopping up holes). I do not have a check list, but have a place for everything and see that everything is in its place. Of course, I also take other things when I need them, e.g. lockslides when I add excluders and additional boxes. A carbolic impregnated cloth (kept in a closed tin) is sometimes useful when shutting up colonies for moving to encourage the bees hanging out at the entrance to move into the hive.

4. Manley once wrote (in BBJ Dec. 1959) "Where our out-apiaries are it would be virtually impossible to use double brood chambers on account of woodpeckers. Some beekeepers, among them Alec Gale, are lucky, and are not much troubled by these birds. They bother me - and plenty of others too."

Chapter 11. Heather Honey (pages 183-196)

1. W.Hamilton, "The Art of Beekeeping", Herald, 1945. Hamilton's book remains, in my view, one of the very best books on practical beekeeping for those who use the 14 x 8½ inch frame. Although long out of print, good secondhand copies are, fortunately, still readily available.
2. J.Ashton, "Colony management for maximum ling heather honey production", Kirkley Hall, March 1966.
3. Neil Anderson, Lenzie, Glasgow.
4. In "Beekeeping at Buckfast Abbey". Page 53.
5. Kierulf, in Norway, established a clear correlation between the yield per hive of heather honey and the aggregation of degrees of temperature above 16C (62F) between 20th June and 20th July. With a total below 30 it was not worth taking bees to the heather, whereas with a total above 60 yields were high even when the period between 20th July and 20th August was comparatively cold.
6. Actually, 370 lbs. of cut comb heather honey from the ten colonies, plus 30 lbs. of honey in the jar.
7. The Norwegian heather honey loosener has individually sprung nylon needles, which plunge into each cell. It is a costly machine, but very efficient. Semi-tangential swing basket extractors are used in conjunction with the Norwegian loosener; a radial extractor won't do the job, however fast it runs. It has been found possible to remove the resultant wax particles from the honey by pumping it from the extractor through a heat exchanger several metres long,

where it is warmed to 110 degrees F (44 degrees C), and then through a spin float separator, to produce a satisfactory wax free product.

8. Except by very sophisticated and costly means. See Note 7 above.

9. In 1956 Brother Adam gave a talk to Devon Beekeepers on taking bees to the moor, getting heather honey, and dealing with the crop. I took notes at the time, which I still have. He made three points that are still relevant and important; first, that "there is no other honey of which the flavour, aroma and consistency is so easily ruined by heating. So do not heat ling honey unless you must, and if you do, take the utmost care, do not exceed 130 degrees F, and do not keep it at that temperature, or even somewhat lower, for any length of time." Second, that "some races and strains of bees are useless at the heather, while others, notably those developed in predominantly heather areas, are very good. In the development of the Buckfast bee, performance at the moor has always been of first importance."; and third, that "there is no satisfactory way by which ling heather honey can be strained. If particles of wax get into the honey they cannot be removed. So pressing heather honey from the comb must be done through clean, strong, linen scrim straining cloth that will not permit even the tiniest particles of wax to pass through.".

10. See Illustration, Fig 27, based on Neil Anderson's drawings; also "Cogs for cut comb", by George Tough, in The Scottish Beekeeper, September 1985, and Morse and Hooper, Encyclopaedia of Beekeeping, page 165.

Chapter 12. Feeding Bees (pages 199-208)

1 2-gallon (10 litre) plastic contact feeders are now available from the appliance manufacturers.

2 A boiling solution of washing soda is more effective than boiling water, but it is nasty stuff that requires careful use. Industrial alcohol would do the job, but only if the gauze could be immersed in it and left for a time for the propolis to soften and dissolve. Perhaps there is a better way. I try to avoid the necessity.

3 The addition of one level teaspoonful of salt per gallon of winter feed syrup is said to be effective in checking or preventing chalk brood. The cost is negligible. Wedmore recommended adding salt to the water source provided for bees to drink at.

4 op cit.

5 Wallace, Quince Honey Farm, regularly feeds pollen patties in spring, and uses the following recipe, viz. 4 lbs. pollen, 12 lbs. defatted soya flour, 21 lbs. sugar, 11 lbs. water. Mix pollen thoroughly with hot water, add sugar and again mix thoroughly,

then add soya flour and mix again. Make into 6 oz. patties and put each between kraft paper.

6 The subject is more fully discussed in Johansson and Johansson's "Some Important Operations in Bee Management". IBRA 1978.

7 Reported in "Gleanings", July 1985.

8 "Rambler", The Scottish Beekeeper, Oct, 1984.

9 In "Honey Production in the British Isles".

10 Dr.L.Bailey, Gooding Memorial Lecture, 1984, "The effect of the number of honeybee colonies on their honey yields and diseases", CABk.

Chapter 13. Etceteras (pages 213-226)

1. Manley also deals with this subject very thoroughly in "Beekeeping in Britain", 1948, which has a chapter on Preparing and Marketing the Crop.

2. The advent of the microwave oven may change all that if a satisfactory technique can be developed to use the microwave to liquefy honey that has granulated in the comb, with the result that it will not re-granulate for some weeks. The production of comb honey from the rape may thus become a practical operation. Except for the certainty of granulation in the comb, the volume and intensity of the nectar flow from the rape, in good weather, certainly provides the opportunity for the production both of cut comb and of sections.

3. Round plastic comb honey sections first appeared in USA in 1954 as Cobana sections. They are now made by more than one manufacturer but are standardised and interchangeable. The bees have no corners to fill (as they have in rectangular sections) and have no access to outer surfaces, which eliminates the need for scraping and cleaning. Clear plastic covers both display the comb to advantage and protect against damage. Filled, a round section contains about 9 ounces of honey. Square sections contain 12 to 14 ounces. The retail price is usually much the same for the two types. Good round sections, containing a little over half a pound of honey, will sell for more than a one pound jar of extracted honey, perhaps 50% more.

4. Richard Taylor "The New Comb Honey Book."

5. I have a sterilizing box in which I can hang up to a hundred combs on two racks. Comb storage in such a box releases brood boxes and supers for cleaning and repair or to cover fondant or pollen patties. Sterilizing in such an airtight container also economises in the sterilizing materials.

6. This method of using Fumidil B is suggested by Edward Crimmins, in The Beekeepers Quarterly No.6. I had been told of it in Canada

somewhat earlier, and have used it in this way since then. The stuff is so expensive that an economical method of using it effectively is well worthwhile.

7. Roger Morse's booklet "Making Mead" (Wicwas Press 1980) is an outline of its history, and of recipes, methods and equipment for making it.

8. As Manley said (in "Beekeeping in Britain") "Remember that the range of your bees from their hives covers an area of five or six thousand acres at least, and then consider how much influence a row of Mignonette is likely to have".

Bibliography

1. A short list of the most useful books must be a personal choice, about which no two beekeepers might agree. However, I have added an asterisk to those books that I would most strongly recommend.

APPENDIX 1

Changes in farming practice in England and Wales and their implications for beekeeping.

Donald Sims, FRICS, NDB.

The keynote address given by Mr Sims to the BBKA Spring Convention at Stoneleigh in April 1983, and published (in slightly abbreviated form) in "Bee World" later that year.

I have kept bees of my own since 1926, and until very recently I spent my life in agriculture. I have run about forty colonies of bees for more than thirty years, on farms as far apart as Kent, Devon, and Northumberland. In Essex and Cambridgeshire I still do. I have thus constantly been made aware of the impact on beekeeping of the changes that have taken place in farming during those years.

In the 1930s I was learning to farm, first in Kent and later in Suffolk. Romney sheep grazed the orchards and pastures of the Weald of Kent, wild white clover was grown for seed, the summers were good, and bees flourished. The sheep in the orchards were replaced by the gang mower, and now the cherry orchards and most of the apples have gone, and the clover seed crops too. I had bees in Suffolk in the 30s when there were fields with a fifteen-year growth of hawthorn and bramble, abandoned as unprofitable all those years before, and not used since. Thistles dominated arable land. The bottom had gone out of farming, and bees thrived on the neglect. We had very good crops of lovely honey from white clover, charlock, thistle, hawthorn and bramble. But we could not sell it, except at a give-away price.

After the Second World War and another spell in Kent, I joined the Ministry of Agriculture, and have since lived and kept

255

bees in Derbyshire, Devon, and Northumberland, and more recently in East Anglia once more.

The pattern of change in farming

Changes in farm practice, changes in farm crops, and changes in the farm scene, all affect beekeeping. Fig 39 shows the broad pattern of change between 1932 and 1982. It shows the significant increase in the proportion of land used for arable crops, with a corresponding decrease in pasture, except in the west (Devon and Wales) where the wartime increase in arable farming has not been sustained. But in these areas, and in the north, the number of cattle and sheep has greatly increased.

The arable areas of eastern England have become more and more arable, while Wales and western and northern England remain mainly pastoral but carry increasing numbers of livestock. The sources of nectar and pollen thus differ greatly from east to west and north. In eastern England they must be found in arable crops or not at all, whereas in Wales and in western and northern England beekeepers look mainly to the grassland.

Changes in farm practice

Two changes are of particular significance for beekeepers; namely the control of weeds, pests, and diseases of farm crops by spraying, and the cutting and harvesting of forage crops at an earlier stage of growth than hitherto. Grass leys with clover, or with lucerne or sainfoin, grown as forage for livestock, have traditionally been cut for hay at full bloom, or a little later, and have thus permitted honeybees to gather a useful crop of honey if the weather was right. They have often been the main source of honey. Such leys are now cut at first flower, or before, for silage, or even for hay, since the value is in the high protein content, and this declines as the crop matures. Two or three cuts a year for silage, with no clover, offer nothing for bees or the beekeeper.

The spraying of farm crops and grassland has become routine practice. Sprays may apply fungicides, herbicides,

256

insecticides, growth regulators, minor and trace elements, and even fertilisers. More than two-thirds of the farm sprays are harmless to honeybees. Many that are toxic rarely come into contact with honeybees, for example sprays used on greenstuffs such as cabbage and brussels sprout crops. But insecticides are intended to kill insects, and the chemicals cannot distinguish between harmful and beneficial insects. The killing of beneficial insects is not new; honeybees were killed by spray poisoning in apple and pear orchards in the 1870s. Indeed, the sprays then used, and the practice of spraying open blossoms, were far more damaging to honeybee colonies in the orchards than modern sprays and modern methods. In both grassland and cereal crops, weeds are now almost eliminated by herbicides, and some of them were excellent sources of nectar and pollen, for example dandelion, charlock, thistle, yellow rocket and polygonum.

The fields growing cereals, wheat and barley, were yellow with charlock in spring; they looked very much like a rather thin patch of oil seed rape does today, and were just as useful to the bees. And the same fields a bit later in the year were a hazy, mauvy, blue with thistles, which produced a delicious water-white honey. Thistles were always present in profusion in these crops; there was no way of eliminating them. Hand hoeing in spring could reduce them, but it was not very effective, and was back breaking work, as I know from personal experience. Thistles in the crop at harvest time were one of the reasons for stooking the sheaves and leaving them to dry out; and those who handled the sheaves always wore gloves, if they had them, for the job. Today it is rare to see thistles, or any other weeds, in the crops; they have all been eliminated by spraying.

By modern standards the yields per acre of wheat and barley were abysmally low and the whole process of growing and harvesting was costly and inefficient. A ton and a quarter of wheat per acre (1200 kg per hectare) was considered a satisfactory yield; there was seldom more and often less. Today yields of three tons per acre of wheat (2700 kg per hectare) are commonplace, and yields of four tons per acre are not uncommon. Yields of wheat, and of barley, have increased threefold in these fifty years. The labour requirement was enormous, and there was nothing for sale until as much as half a year after harvest. Labour cost little per man hour, but the number of man hours needed merely to get in the crop at harvest time was about twenty man hours per ton by the time the crop was in stack, and more to come at threshing time. Today, here in Foxton, and no doubt elsewhere, a combine harvester supported by two tractors with trailers will clear thirty acres in a day if the weather is good, and the grower will have eighty tons or more of wheat on the barn floor ready for sale, or sold and away to the miller by the end of the day. That is less than half a man hour per ton. Those who oppose crop spraying for weed control and other modern methods can have neither knowledge nor experience of how things were in the past.

With few exceptions there is no reason nowadays to apply an insecticide to crops in bloom or a herbicide to weeds in flower. But some crops present difficulties in that they flower, and are attractive to bees, over a long period. Field beans and lucerne (alfalfa) are examples; raspberries are another. And if no other source of nectar and pollen is available, honeybees may still be active in winter rape when only a few flowers remain - just when the crop is at the right stage for post-blossom spray control of seed weevil.

Changes in farm crops

The principal change for beekeeping, and a dramatic change from any point of view, is the rapid spread throughout England of rape grown for oilseed. This has, in recent years, increased the honey crop for beekeepers in the areas affected, and it has become almost the sole source of honey for some. The continuing increase in area and the countrywide spread provide a great opportunity for beekeepers to increase honey production, but they also bring problems, both in colony management and in handling the honey crop. The management problem is akin to that faced by beekeepers in the intensive orchard areas of Kent, where I used to keep my bees, and where the principal nectar flow had ended before June had arrived. The honey handling problem is caused by the rapid granulation of rape honey, which will set hard in combs in the hive, and in the extractor.

Another plus factor is the increasing area of field beans, which at one time had almost disappeared as a farm crop. Field beans are a valuable source of protein for animal feed, and they are grown more now because plant breeding has led to varieties that crop more regularly and give greater yields than formerly. There has also been an increase and spread of pick-your-own enterprises which grow raspberries, black and red currants, and other soft fruit, of use to bees, in areas and localities where they would not otherwise be found.

The minus factor of greatest significance for beekeepers is the combined reduction in the area of permanent pasture, and the virtual disappearance of wild white clover from grass seed mixtures used for short term pasture and for fields intended for hay or silage. Wild white clover is no longer a principal source of nectar - and of honey of supreme quality - in eastern England. And the production of clover seed, in Kent and elsewhere, has virtually ceased. However, we now see sheep where formerly we saw none, and one can hope that the recent rapid increase in their numbers may encourage white clover once more. It is an indigenous plant that will make a volunteer appearance in well grazed pastures.

The reduction in the area devoted to orchards, and particularly the disappearance of cherry orchards, is also a minus factor. Large orchard trees have gone because of the difficulty and cost of spraying and picking; there is no dwarfing stock for cherries as yet, so the old orchards have not been replaced. Apples are now grown on dwarfing stocks so that they are easy to spray and to pick, and bees are still necessary for pollination, so that beekeepers have not lost much.

Oilseed rape

Rape seed and mustard seed have been cultivated for centuries for their oil, since before 1500 BC in India and since the thirteenth century in Europe. In the seventeenth century sufficient seed was produced in the UK to meet home demand and leave a surplus for export. Rape seed oil was used mainly as a lamp oil, for which purpose its slow and smoke free burning properties make it very suitable. I guess Aladdin`s lamp burned rape seed oil. It has had a more recent use as an industrial lubricant, for which it is also very suitable. The shortage of edible oils in the Second World War led to the use of rape seed oil instead of imported oils for the manufacture of margarine, and the area of rape grown for seed increased fourfold. Worldwide, rape seed is currently in fifth position as a source of edible oil, after soya bean, sunflower seed, cotton seed and ground nut, and in western Europe it is the most important source. Canada, now the largest exporter of rape seed, first grew rape for oilseed in 1942, principally to supply lubricants for marine and aircraft engines.

Oilseed rape is grown as a break crop in cereal rotations, especially where potatoes or sugar beet are not feasible. It does well in the main areas of cereal production and is neither susceptible to nor a carrier of cereal diseases. Until recently it has not contributed much to farm income, but because of improved varieties, improved techniques, and a stable price for the product, it now does so, and has the great advantages that it can be harvested by combine, and that it ripens before most cereal crops.

In England and Wales the area of oilseed rape increased from 39,000 ha in 1975 to 174,000 ha in 1982 (See Fig 40) and in 1983 is believed to be about 200,000 ha. The area of rape grown for seed now exceeds that of vegetable and potatoes, and now ranks third to wheat and barley. There has been a fourteen-fold increase in area in ten years, a rate of increase of about 30% a year. A continuing increase is highly probable, although there will be an upper limit in due course. But it is in the arable areas, particularly in eastern England, that this dramatic change has taken place, as Fig 40 shows. Two-thirds of all the rape grown for oilseed is in the Eastern and East Midland Regions of the Ministry of Agriculture, Fisheries and Food.

Changes in the farm scene

The removal of hedges has transformed the appearance of some areas of the countryside, particularly in the arable areas of eastern England. It has not much reduced bee forage, since the contribution of hedges was rarely very large. Nor is the resulting open landscape necessarily less pleasing. Large numbers of trees have been planted - certainly more than have

260

been removed - but this will not become evident until they become prominent in the landscape. Such tree planting, in the corners of fields, or to create small woods, will create numerous modest areas where wild flowers will be allowed to grow and provide some forage for bees. This is unlikely to have much significance for beekeeping, except perhaps in providing a suitable site for an out apiary. More significant is the replacement of deal elms by other trees, mainly deciduous species, of which alder, willow, rowan , maple, and horse chestnut will all provide bee forage in due course. The wild flowers on unsprayed and uncut motorway and other road verges , and the widespread planting of ornamental trees and shrubs in gardens and public places, already make a useful contribution to bee forage.

The implications for beekeeping

Changes in agriculture and the countryside are much more evident, and have had much more effect upon beekeepers and beekeeping, in the arable areas of eastern England than elsewhere. I live at Foxton, near Cambridge, on chalk. There were nine beekeepers in the village in 1947 with about a hundred and twenty colonies of bees between them, and by all accounts they did well. In the parish of Foxton at that time there were eighty hectares of permanent grass grazed by cattle and sheep and nearly eighty hectares more of clover and sainfoin leys for mowing. Today there are less than six hectares of permanent grass, of which all but one hectare are in the village recreation ground and a small caravan site, and there are no leys for mowing and no grazing livestock at all. The village population has doubled, but I am the only beekeeper; the two colonies in my garden are the only bees in the parish, and they barely survive. And although there is rape in profusion in neighbouring parishes within a mile or two, on heavier land, sugarbeet and potatoes continue to take preference to rape in Foxton.

The changes in farm practice and farm crops make migratory beekeeping essential in the areas most affected. There are real opportunities and a reasonable expectation of an increased crop of honey if, as beekeepers, we are sufficiently adaptable. Oilseed rape already provides a significant and increasing proportion of the honey crop. Its potential is very great, but in England it is unlikely to be fully realised because it flowers too early in the year. When it is spring sown, as it is in Canada and Sweden, it flowers in July and fits well with colony development. In England, rape grown for oilseed is autumn sown because crop yields are so much greater than from spring sowing, and it flowers in late April into May. Colonies are not at their peak; they develop rapidly on the abundant nectar and pollen, and then tend to swarm, and may starve in June/July. A later source of nectar and pollen is needed, and unless wild white clover becomes sufficiently re-established, which I consider unlikely in eastern England, or one or more of the alternative crops grown for oil, such as borage and evening primrose, become widespread, which I also think unlikely, there is only heather, in August, many miles away from most of the rape. But the need to move our bees to farm crops that produce nectar has become evident.

Migratory beekeeping presents new problems and requirements for most beekeepers, which lie outside the scope of this paper. But all of us who keep our bees in rural areas have to face the dramatic changes in farming, and in the countryside, that I have tried to tell you of today, and adapt our practice and methods accordingly.

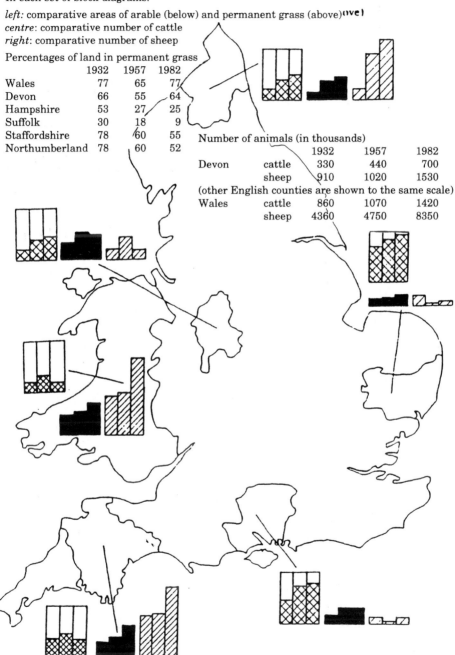

In each set of block diagrams:

left: comparative areas of arable (below) and permanent grass (above)**(ve)**
centre: comparative number of cattle
right: comparative number of sheep

Percentages of land in permanent grass

	1932	1957	1982
Wales	77	65	77
Devon	66	55	64
Hampshire	53	27	25
Suffolk	30	18	9
Staffordshire	78	60	55
Northumberland	78	60	52

Number of animals (in thousands)

		1932	1957	1982
Devon	cattle	330	440	700
	sheep	910	1020	1530

(other English counties are shown to the same scale)

Wales	cattle	860	1070	1420
	sheep	4360	4750	8350

Fig 39   Grass and arable at 25-year intervals

Proportion of permanent grass to arable land at 25-year intervals (1932, 1957, 1982) in Wales (left) and in 5 representative English counties.   263

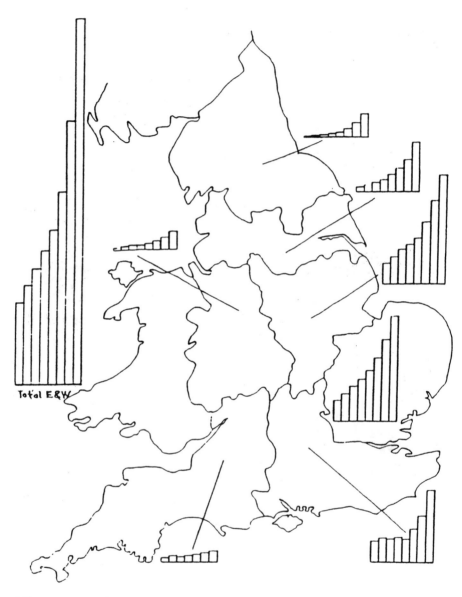

Total E&W

Fig 40   Oilseed rape in England and Wales

Areas of oilseed rape in the 8 Ministry of Agriculture regions of England and Wales 1975 - 1982.

The total areas (in thousand hectares) are:

1975	1976	1977	1978	1979	1980	1981	1982
39	47	53	64	74	92	125	174

APPENDIX 2

Making Mead

Mead is an alcoholic drink made by fermenting honey and water with yeast. Just as honey was for thousands of years the principal sweetening agent known to mankind, so mead was probably the first alcoholic drink known to man. Honey is markedly hygroscopic, that is it readily absorbs moisture, and the addition of water to honey automatically produces a mixture which can be fermented by yeast. Wild yeasts are commonly present in the atmosphere and honey exposed to moist air will eventually ferment. It is highly probable that the discovery of alcohol, or at least of an alcoholic drink, occurred in this accidental manner.

Certainly mead was a principal alcoholic drink of most if not all early civilisations, and it continued to be so in many parts of the world until well into the eighteenth century, when the production of sugar from cane in the West Indies, and later from beet in Europe, provided both an alternative and eventually a cheaper sweetening agent an basis for alcoholic drinks. Honey was for long the common sweetener used in brewing ale as well as for making wine, and in combination with malt, fruit juices, and spices, to produce an astonishing variety of drinks, some with curious and romantic names and many with allegedly aphrodisiacal qualities.

Any alcoholic drink made principally from honey is commonly called "mead", but it is convenient and more accurate to distinguish five groups of such drinks, namely mead, melomel, metheglin, pyment and cyser. Metheglins are made by fermenting honey and spices or by steeping spices in mead. Melomels are fermented honey and fruit juice. They retain much of the character of mead but use much less honey and are thus less expensive. They can properly be considered as light white wines, and most fruits with light coloured juices can be used, as can canned juices such as orange, grapefruit and pineapple. The qualityof the honey is also less important than for meads. Pyments are made by fermenting grape juice with added honey. They originated in Greek and Roman times, when it was found that the addition of honey to the grape juice increased the alcoholic strength of the wine. Cysers are made by fermenting apple juice with added honey, and should be a stronger drink than cider, with added flavour.

Mead is made by fermenting honey with yeast, nothing being added other than the yeast nutrients. The quality and character of the honey is therefore of first importance. A good mead is very much like a good white wine and should have a delicate flavour varying with the source of the honey used. A mead made from a pale honey, such as would be acceptable in the class for light honeys at a honey show, will be very different from one made from a dark honey or from heather honey.

Typically, meads made from pale, light honey of good quality and flavour, say from wild white clover, will produce a pale coloured mead of delicate flavour, best if slightly sweet. Such a light, sweet mead can be made and drunk in six weeks or not much more, and can be quite fragrant and delicious. At the other extreme, meads made from heather honey mature slowly and are not likely to be ready for drinking for a year or two and may be at their best only after several years.

Mead can be, and often is, made by beekeepers from the watery liquid in which wax cappings have been soaked after extracting honey, or from soaking in water the wax cake left from the heather honey press. Both may give good meads, if the beekeeper knows how and has good judgement of honey quality, but my general advice to beekeepers would be to add sugar to this dilute solution of honey in water and feed it to the bees. Use some of your very best honey to make mead and the result will justify your choice if you make it well.

Recipes for making mead are to be found in many of the bee books, and there have been some notable pioneers of good mead-making in the beekeeping fraternity, notably Brother Adam at Buckfast.

Wine making at home has become very popular in recent years. Grape and other fruit concentrates are readily available, as are the necessary items of equipment. Pleasant and enjoyable wines are easily made by simple methods that are nevertheless thoroughly well based and well tried. The techniques of home wine-making have both been simplified and improved and made available to all.

There are probably many more home wine-makers than there are beekeepers, and some of the best amateur wine-makers have turned their attention to mead-making and to the use of honey in making wines, and the best advice and instruction on the subject is now to be found in amateur wine-making publications. Those who intend to make mead should read "First Steps in Winemaking" by C J Barry and "Making Mead" by Bryan Acton and Peter Duncan. Both are "Amateur Winemaker" publications, obtainable at or through home wine-making stores.

Basic equipment for wine or mead making is simple and inexpensive. You will need a fermentation vessel, preferably two to facilitate racking, that is siphoning to leave the lees (sediment) behind when fermentation ceases, and for storage before bottling. The standard one gallon (5 litre) glass demijohn is ideal as the necessary bung and air-lock to fit the narrow neck is available wherever wine-making equipment is sold. An hydrometer is a great help. A dry mead results from a honey solution with a specific gravity of 1.075 and will have close to 10% of alcohol. A sweet mead needs more honey and a solution with a specific gravity of 1.115.

Every piece of equipment should be sterilized before use in wine and mead making, and this is best achieved by washing or immersing in a sterilizing solution containing sulphur dioxide, easily prepared from sodium metabisulphite, and a 10% stock solution will keep for months in a sealed bottle and can be re-used many times. A proprietary form of sulphite known as Camden tablets can be similarly used to make a sterilizing solution or to inhibit fermentation. In mead making the honey and water mixture is commonly boiled to sterilize it, and the yeast added when the mixture is sufficiently cool. A better mead is likely to result if the honey and water mixture is sterilized with sulphite (10 ml of stock solution or two Camden tablets per gallon) and the yeast add twenty-four hours later.

Neither light nor dark honey contains any acid or tannin to speak of, and since both are necessary to the full development of wine or mead, they have to be added. Honeydoes not contain any natural yeast nutrient either, and a sufficient quantity of nutrient has also to be added to ensure a satisfactory fermentation of honey. Yeast nutrient is readily available from chemists and others supplying amateur wine-makers' requirements, both as tablets and as loose crystals. It is principally ammonium phosphate, which at ½ tsp per gallon (or 5 litres) will provide the nitrogen and phosphorus essential for a strong fermentation. It is probably better to use a proprietary yeast nutrient at one tsp (or two tablets) per gallon (5 litres).

Acids also play an essential part in fermentation, but the amount required is very small. The addition of citric acid (one oz per gallon) is traditional, and many will like the flavour it imparts, but a combination of malic acid and tartaric acid in a two to one ratio (½ oz malic acid and ¼ oz tartaric acid per gallon) will permit a superior mead to be developed.

A tiny quantity of tannin in the form either of about 1/15 oz tannic acid or a few drops of wine tannin per gallon is also needed. Probably the most convenient way of providing such small quantities is to get the chemist to make up a solution of 10g of tannic acid, 35g of tartaric acid, and 70g of malic acid, in 250ml (or half a pint) of distilled water, and then measure out 2 fl oz (or 50ml) of the solution to a gallon (5 litres). The remaining solution will keep in a bottle with a tight fitting screw lid in a refrigerator for use with a further four gallons. Or make five gallons of mead and use the lot.

Maury yeast is widely recommended for mead making and is an excellent yeast for this purpose, particularly with heather honey. But it s a rather slow fermenting yeast that produces flocculent sediment that is easily disturbed and makes racking difficult. Sedimentary wine yeasts make excellent meads and are readily available, and a Rhine yeast for a dry mead and a Sauternes or Sherry yeast for a sweet mead will be very suitable.

Wine yeasts should be activated before use to ensure an active ferment with no delay. To half a pint (250 ml) of cold, boiled, water add a dessertspoon full of sugar, an ounce (25 ml) of orange juice or lemon juice, a little yeast nutrient and the yeast of your choice. Plug the neck of the bottle with cotton wool. Keep the bottle in a warm place and shake gently from time to time and it will be ready for use in forty-eight hours.

The basic procedure is to add the yeast nutrient and acid mixture to half a gallon of warm water and stir in honey sufficient for one gallon until the liquid is of uniform consistency, then make up to one gallon with cold water and add 10 ml of the sulphite solution. Twenty-four hours later add the activated yeast, and ferment to dryness in a warm place (70 to 75 degrees F). Rack within a week of fermentation ceasing, and after three months rack again. Six months later bottle and mature for three months (at least) or more.

Serve mead cool and preferably chilled, either in place of a white wine at table or as an occasional drink with a biscuit. Don't aim at a high alcoholic content; the delicate flavour will probably get lost. Just use a larger glass.

Mead is something of an acquired taste, and not to everyone's liking. If you don't like the mead you have made, blend it with a fruit wine and you will find the resulting wine very palatable.

Melomels, which are fermented honey and fruit juice, fruit meads if you wish, are usually quite delightful. They are much less expensive than meads, ferment more speedily and mature quickly, so they are ready to drink sooner. A can of light fruit juice fermented with 2 lb (1 kg) of rape honey and a Rhine yeast will make six bottles of a pleasant and enjoyable wine in three months, or less if you are in a hurry. Serve it well chilled and in a long glass.

Within half a mile of my apiaries
stands this massive oak tree, more than 700 years old,
and 29 ft. (8.8m) in girth
It is the largest in Essex and the 5ᵗʰ largest in the UK

270

Fisk, Jim, 7
flat combs, 64
flies, 20
flower honey, 137
follower board. *See* dummy
   board
followiness, 240
fondant, 114, 200, 203, 206, 207
foraging bees, 116
Forestry Commission, 187, 234
Fortnum & Mason, 1
foundation, 9, 138, 189
   in supers, 108
   starters, 26, 195
   thin, 218
Foxton, 6, 16, 234, 258, 261
frame
   14" x 12", 138
   bottom bars, 58, 62, 120
   British Deep, 60
   British Standard, 21, 60
   cleaning, 221
   Dadant, 62, 138
   depth, 59
   feeder, 68, 86
   Hoffman, 10, 12, 37, 57, 58, 61,
      62, 63, 243
   Langstroth, 59, 62, 138, 242
   Manley, 77, 80, 195, 218
      4 inch deep, 218
   size, 59
   spacing, 63
   top bars, 58, 61, 62, 120
framed excluders, 70
Free, John, 249
frosting, 216
fruit blossom, 11
frustration methods, 145
fuel, 72, 177
fumagillin, 222
fume board, 78, 83, 118
Fumidil B, 195, 221, 253
fumigate combs with acetic
   acid, 221

G

Gale, Alec, 236
gardening, 2
Gleanings in Bee Culture, 249,
   253
gloves, 48, 53, 81, 82, 177
good tempered bees, 115
Granton knife, 80
granulated honey, 164
granulation, 161, 164, 165, 186,
   217, 258
   solid, 166
   with a fine grain, 214
Gregg, Dr A L, 144, 151, 241,
   248, 249
grooved side bars, 63, 218
grouse, 186
Guilfoyle, John, 59

H

half-size Dadant combs, 112
Hamilton, William, 183, 251
hardware cloth, 70
Hasting, J E, 19, 237
hawthorn, 7, 255
heat shield, 72
heather, 134, 136, 180, 262
   moorland, 186
   moors, 14, 15, 109, 111, 179
   spraying, 187
heather honey, 11, 13, 94, 103,
   183, 184, 185, 194, 218
   crop, 96
   cut comb, 184
   extracting, 191
   in the comb, 184
   ling, 185, 191, 192, 193, 195
   not heather honey, 185
   scraping, 81
   yields from, 188, 251
Heathfield, 234
heavy lifting, 12
hedges
   removal of, 260
Herefordshire, 7

Herzog, 77
Hexham, 14, 130, 179
Hexham apiary, 105
hive
    14x12, 235
    British Commercial, 248
    British Deep, 21, 241, 248
    Commercial, 7, 241
    Dadant, 12, 59, 194, 235, 242, 247
    Langstroth, 2, 5, 7, 10, 57, 207, 234, 235, 241, 242
    Langstroth Jumbo, 21, 59, 60, 235
    National, 8, 12, 15, 19, 58, 60, 61, 116, 235, 241, 243, 248
    Simmins Conqueror, 8
    single walled, 12
    Smith, 8, 12, 15, 57, 58, 60, 63, 116, 138, 184, 235, 241, 242, 243, 247
    stands, 29, 177
    staples, 112
    straps, 112
    tool, 80
    tools, 79
        J-type, 79, 80
        Root type, 72, 80
    two chamber, 242
    WBC, *8*, 58, 241
    Yarrow, 69, 237
Hoffman frame, 10, 12, 37, 57, 58, 61, 62, 63, 243
Holland, Jim, 248
holly, 11
honey
    apple blossom, 11, 13
    bottling, 216
    comb, 6, 190
    flower, 15, 137
    granulated, 164
    granulated in the comb, 165
    hawthorn, 13
    heather. *See* heather honey
    melting it down, 166
    rape flower, 171
    seeding, 215
    to be blended, 215

    turnip, 10
honey extractors
    MG Parallel Radial, 77, 236
    radial, 78
    swing-basket, 192
    tangential, 191
Honey Farming, 79
Hooper, Ted, 17, 169, 252
Howes, Dr F N, 158, 249
Hubbell, Sue, 246
Hunt, John, 9, 12, 100, 219, 234, 235, 236, 249

I

Illingworth, Leonard, 6, 7, 233
inner cover, 71, 85
inspection, 107
    purpose, 51, 107
    routine, 49, 106, 120
    time, 12
interchangeability, 177
interchangeable combs, 64
Isle of Wight disease, 5
Italian bees, 63
Italian queens, 236

J

Jay, S C, 246
Jekyll, Gertrude, 2
Jenner, George, 13, 112
Johansson, T S K & M F, 205, 243, 250, 253
Journal of Apicultural Research, 246
J-type hive tool. *See* hive, tool
Judge, George, 5, 233

K

Kemble, Peter, 239
Kent, 2, 5, 7, *8*, 130, 184, 231, 233, 236, 255
    Weald of, 255
Kent BKA, 233
Kent wild white clover, 235

Scottish BKA, 16
Scottish Colleges, 183
Scraping heather honey, 81
scraping the comb, 165
screen board, 111, 131, 132
secure entrance block, 110
seed crops, 109
seeding honey, 215
settling tank, 215
shaken swarm, 119, 155, 162
shaking bees, 37, 53, 54, 83, 136
shook swarming, 150, 151
shorthand, 88, 89, 115
Simmins, Samuel, 7, 203, 204, 234, 235
Sims, R P, 9, 235
single box colony, 160
single handed, 1, 14
sister queens, 20
Skinner & Rook, 184
slatted rack, 64, 68
Smith, 57, 58, 60, 61, 63, 102, 120, 184, 235, 241, 248
    9 comb, 237
    brood box, 113, 137
    colony, 159
    cutter/scraper, 81
    deep boxes, 139
    hive, 116, 138, 242, 243, 247
    ten comb, 134
smoker, 48, 50, 71, 177
    Dadant, 72
    fuel, 81
    Root, 72
    Woodman, 6, 233
Snelgrove system, 134
snowdrops, 175, 180
Soden, Leslie, 236
Soft fruit, 109
soft string, 111
South Downs, 11, 101
Spain, 250
spare time activity, 2
spin drier, 165, 192
split board, 84, 136, 146, 178

split in two, 103
spray damage, 161, 163, 164
spray poisoning, 257
sprayguard entrance closures, 164
spraying, 235, 256
spraying heather, 187
Spring division, 116, 134
spruce honeydew, 20
squeezing through linen bag, 165
standardise, 17
starters, 166, 184, 189, 218
sterilizing box, 253
Stock, Wally, 234
Stoneleigh, 15
stored combs
    check, 221
strain of bee, 12, 48, 99, 101, 102, 115, 140, 141, 143, 145, 161, 189, 236, 238
Strainaway warming and straining outfit, 250
straw skeps, 8
Sturges, A M, 234, 235, 236
Suffolk, 7, 235, 255
super of drawn comb, 118
super of foundation, 118
supers, 77, 103
    too late, 143
supersedure, 114
Sussex, 129, 130, 234, 235, 242
swarm
    artificial, 23, 137, 145, 146, 151
    control, 116, 140, 142, 161
    control methods, 145
    natural, 149, 151, 153, 154
    queen, 155
    shaken, 119, 155
    shook, 145, 147, 150, 151
    taking, 54
    Taranov, 148, 149
    using, 153
swarming, 101, 118, 161
Sweden, 158, 262
sweet chestnut, 11, 100